LANCHESTER LIBRARY

3 8001 00114 3605

.ester Library

£22.00

Geology of the country around Redditch

The district described in this memoir includes the towns of Redditch and Bromsgrove, the southern suburbs of Birmingham and the picturesque Forest of Arden. In terms of geological structure, the western part of the district lies within the Worcester Basin and is separated from the Knowle Basin farther east by the uplifted Lickey Ridge. The Lickey Hills are geologically complex and include the outcrops of the Barnt Green Volcanic Formation and the Lickey Quartzite, both of possible Tremadocian age, as well as those of several Silurian formations. In the east, known Tremadocian rocks are proved at depth but do not come to crop.

On the flanks of the Lickey Ridge, the Carboniferous succession includes the Halesowen, Keele and Bowhills formations. In the east, within the Warwickshire Coalfield, much more of the sequence is proved, including the Lower and Middle Coal Measures and the Etruria Marl. The Thick Coal within the Middle Coal Measures is of great economic importance. Above the Halesowen Formation, the Carboniferous rocks are red, poorly fossiliferous, continental deposits.

The Permian Clent Breccias are coarse clastic deposits derived largely from local Lower Palaeozoic rocks. This structural inversion, whereby an upstanding area was changed into a sedimentary basin bounded by growth faults, occurred during Triassic times. Indeed, both the Worcester and Knowle basins contain thick sequences of Triassic rocks and are bounded on their eastern margins by large growth faults, throwing down to the west. The type areas for the relatively fossiliferous Bromsgrove and Arden sandstones occur in the Worcester and Knowle basins respectively.

An extensive marine transgression towards the end of Triassic times is represented by the richly fossiliferous mudstones and limestones of the Penarth Group; similar rocks persist throughout the Jurassic succession.

Much of the district is drift-free, but there are extensive spreads of glacial deposits in the north, and younger river terrace deposits are widespread, especially in the south.

The memoir also describes the tectonic history, mineral products and hydrogeology of the district.

D1341666

Cover photograph
Wootton Wawen church, constructed from locally quarried Arden Sandstone (A 13750)

Plate 1 West-facing escarpment of the Arden Sandstone north of Beaudesert, Henley-in-Arden (A13489)

BRITISH GEOLOGICAL SURVEY

R A OLD,
R J O HAMBLIN,
K AMBROSE and
G WARRINGTON

Geology of the country around Redditch

Memoir for 1:50 000 geological sheet 183
(England and Wales)

Contributors

Geophysics
R M Carruthers

Hydrogeology
R A Monkhouse

Palaeontology
H C Ivimey-Cook
S G Molyneux
B Owens
A W A Rushton

Petrology
G E Strong

LONDON: HMSO 1991

iv

© *NERC copyright 1991*

First published 1991

ISBN 0 11 884477 6

Bibliographical reference

OLD, R A, HAMBLIN, R J O, AMBROSE, K, and WARRINGTON, G. 1991.   Geology of the country around Redditch.   *Memoir of the British Geological Survey*, Sheet 183 (England and Wales).

Authors

R A Old, BSc, PhD
R J O Hamblin, BSc, PhD
K Ambrose, BSc
G Warrington, BSc, PhD
British Geological Survey, Keyworth

Contributors

R M Carruthers, BSc, H C Ivimey-Cook, BSc, PhD,
S G Molyneux, BSc, PhD, B Owens, BSc, PhD,
A W A Rushton, BSc, PhD, G E Strong, BSc
British Geological Survey, Keyworth

R A Monkhouse
British Geological Survey, Wallingford

P 0 2 9 7 1

Other publications of the Survey dealing with this district and adjoining districts

BOOKS

Memoirs
Dudley and Bridgnorth, sheet 167
Geology of the country around Droitwich, Abberley and Kidderminster, sheet 182
Geology of the country around Warwick, sheet 184
Geology of the country around Stratford-upon-Avon, sheet 200
Geology of the country around Banbury and Edge Hill, sheet 201

British Regional Geology
Central England, 3rd edition, 1969

MAPS

1:1 000 000
Pre-Permian geology
Solid (pre-Quaternary) geology (south sheet)

1:625 000
Solid geology (south sheet)
Quaternary geology (south sheet)
Aeromagnetic map (south sheet)

1:250 000
Solid geology, East Midlands
Aeromagnetic anomaly, East Midlands
Bouguer gravity anomaly, East Midlands

1:63 360 and 1:50 000
Sheet 167   Dudley (Solid and Drift)
Sheet 168   Birmingham (Solid, Drift)
Sheet 169   Coventry (Solid and Drift)
Sheet 182   Droitwich (Solid and Drift)
Sheet 183   Redditch (Solid and Drift)
Sheet 184   Warwick (Solid and Drift)
Sheet 200   Stratford-upon-Avon (Solid and Drift)
Sheet 201   Banbury (Solid and Drift)

Printed in the United Kingdom for HMSO
Dd 291129   C10   8/91

LANCHESTER LIBRARY

CONTENTS

FIGURES

vi

TABLES

PLATES

PREFACE

This memoir describes the geology of the district covered by the Redditch (183) 1:50 000 sheet, published in 1989, and is intended to be read in conjunction with the map. The district includes the town of Redditch and the southernmost suburbs of Birmingham. In common with all urban areas, these towns require considerable quantities of sand, gravel and crushed rock aggregate to support the local construction industry and ideally sources should be close at hand. The geological map and accompanying memoir provide the basic information needed by industry to locate these supplies. In addition, the 1:10 000-scale maps, on which this information is based (listed in Appendix 1), are important to both the planning authorities and the construction industry in providing indications of areas at risk from landslips, slope instability and, a special factor in this district, cavities formed by the solution of gypsum and limestone in the underlying bedrock.

Part of the local water supply is derived from groundwater sources and there is a long history of mineral extraction in the district, including coal, lime, gypsum and salt. Each has created its own problems, particularly coal mining and brine pumping, which can be assessed only by reference to the geological maps.

There are three important sedimentary basins in the area; the western edge of the concealed Warwickshire Coalfield in the east, and the Knowle and Worcester basins containing Triassic rocks. In the north-west of the sheet area are the Lickey Hills where, in a structurally complex zone, Lower Palaeozoic rocks have been upfaulted along a ridge between the Knowle and Worcester basins. The glacial deposits in the district give important information on the last glacial episode. There are two distinctive suites of deposits: one from westerly, the other from easterly derived ice sheets, and an area of overlap is defined.

Until the recent remapping, there had been no detailed survey of the district since the original primary mapping at one inch to one mile by H H Howell, E Hull, J B Jukes and A C Ramsay, published as Old Series one-inch sheets 54NW, 54NE, 54SW and 54SE in 1854 and 1855. After that, only small areas along the northern and western margins were surveyed on the scale of six inches to one mile during the present century as overlap from the mapping of the adjoining one-inch sheets 168 (Birmingham) and 182 (Droitwich). The recent resurvey, therefore, represents a major advance in the understanding of the geological history of the district.

Peter J Cook, DSc
Director

British Geological Survey
Keyworth
Nottingham
NG12 5GG

12 April 1991

NOTES

ACKNOWLEDGEMENTS

Throughout the memoir the word 'district' refers to the area covered by the 1:50 000 geological sheet 183 (Redditch).

National Grid references are given in square brackets; those beginning with the figure 9 lie in the 100 km square SO and those beginning with the figures 0, 1 or 2 lie in the 100 km square SP.

Numbers preceded by the letter E refer to the sliced rock collection of the British Geological Survey.

The area of Redditch New Town was surveyed on the six-inch scale by A Horton and B C Worssam between 1969 and 1972, and the remainder of the district was mapped on the 1:50 000 scale by K Ambrose, P J Strange, R J O Hamblin, R A Old and A A Jackson, between 1979 and 1982. The latter survey was supported largely by the Department of the Environment to provide information for land use planning, especially with respect to potential resources of sand and gravel. The sections on the Ordovician, Quaternary, structure and economic geology were written by Dr Old, whilst the Silurian, Carboniferous and Permian sections were compiled by Dr Hamblin. The Triassic section is by Drs Old, Warrington and Hamblin, and the Jurassic account is by Mr Ambrose. The survey was under the supervision of Dr G A Kellaway and Mr G W Green as District Geologists, and Dr A J Wadge as Regional Geologist.

Contributions to the palaeontology were made by Drs H C Ivimey-Cook, B Owens, S G Molyneux, A W A Rushton, G Warrington and D E White. Petrological descriptions of thin sections of various rocks were made by Mr G E Strong. Mr R M Carruthers has interpreted the results of geophysical investigations and Mr R A Monkhouse has contributed the section on water supply. The memoir was edited by Dr A J Wadge.

Grateful acknowledgement is made to numerous organisations and individuals who supplied data during the survey; landowners, quarry operators and local authorities all gave generous help to our task. Particular mention is made of the British Coal Corporation and the Redditch Development Corporation in this regard, and of the Warwickshire Museum for the loan of specimens.

ONE

Introduction

The district around Redditch (Sheet 183), described in this memoir, lies to the south of Birmingham, in the counties of Warwickshire and Hereford & Worcester. The major centres of population are Redditch, Bromsgrove and the southern suburbs of Birmingham. Most of the remainder of the district is given over to agriculture, with the Lickey Hills Country Park forming an important recreational area.

The dominant topographical features of the district are the valleys of the rivers Alne and Avon, and the principal watershed along the Lickey Hills and The Ridgeway. To the west of this divide, the drainage is predominantly westward, and to the east predominantly southward. The drift-free areas have a more varied topography, typically with low, steep scarps and gentle dip slopes. The drift-covered areas, most extensive in the north and east, have generally low relief.

GEOLOGICAL SEQUENCE

The succession of rocks occurring in the district is shown on the inside front cover. The solid rocks range in age from Cambrian to Jurassic; their distribution is shown in Figure 1. Several of the rock groups are separated by unconformities representing periods during which uplift and erosion took place. The Lower Palaeozoic rocks are restricted at outcrop to the Lickey Hills; a more detailed 1:25 000 map of this area is shown on the 1:50 000 sheet.

GEOLOGICAL HISTORY

A glimpse of the Lower Palaeozoic history of the west of the district is afforded by the pyroclastic rocks and tuffaceous sedimentary rocks of the Barnt Green Volcanic Formation, which crops out in a small fault-bounded inlier. These rocks, of possible Tremadoc age, may pass upwards into the sparsely fossiliferous Lickey Quartzite which is imprecisely dated within the Ordovician. In the east of the district, dark grey mudstones of Tremadoc age (Merevale Shales) have been proved in one borehole, and seismic evidence suggests that a considerable thickness of Cambrian strata underlies at least the eastern part of the district. These strata are likely to belong to the Stockingford Shales, which contain a graptolite and brachiopod fauna and were probably deposited in a shelf sea.

One of several periods of uplift along the north–south Lickey Ridge (see p.54) occurred between the deposition of the Lickey Quartzite and the unconformably overlying Rubery Sandstone and Rubery Shale. These Silurian strata contain a fauna of brachiopods, trilobites and graptolites which indicates a late Llandovery age. Mudstones of early Wenlock age have been proved in boreholes, and brachiopod-rich limestones and mudstones of late Wenlock age crop out in a faulted inlier at Kendal End.

Post-Silurian (Caledonian) earth movements produced uplift and prolonged erosion throughout the district. In the east, erosion laid bare the Merevale Shales before sedimentation resumed in late Carboniferous times. The Productive Coal Measures, of Westphalian A, B and basal C age in the Warwickshire Coalfield, are delta-swamp or estuarine sediments, comprising grey mudstones and siltstones with a few seatearths and sandstones, and an economically important seam, the Thick Coal. The overlying Etruria Marl was deposited in a similar deltaic environment but, because oxidising conditions prevailed, the beds of mudstone and sandstone are mainly red-mottled and variegated, with a few grey horizons and no coals. A return, in Westphalian D times, to a reducing delta-swamp environment occurred during the deposition of the grey sandstones, mudstones and thin sulphurous coals of the Halesowen Formation. In the Lickey Hills area, these are the youngest Carboniferous rocks present and they overstep the Silurian onto the Ordovician beds.

The Keele Formation succeeds conformably and is of continental aspect, comprising barren red and orange mudstones and red sandstones and mudstone-pellet breccias; rare thin limestones contain the worm *Spirorbis*. The Enville Group succeeds conformably and is of similar facies. The Bowhills and Coventry Sandstone formations consist of interbedded, red and green sandstone and red mudstone, whilst the Tile Hill Mudstone sequence contains less sandstone. All three formations are tentatively regarded as Westphalian D in age, but stratigraphically significant fossils are lacking from this part of the succession.

There was renewed uplift in the west of the district during the late Carboniferous, resulting in a strong unconformity at the base of the Clent Breccias. The clasts in the breccias include sandstone, shale and abundant volcanic rocks; the last may come from the Barnt Green Volcanic Formation or local correlatives. The breccias were laid down as fans of coarse debris, banked against higher ground to the south of their present outcrops. They contain no stratigraphically diagnostic fossils, so their age is uncertain, but they are unconformably overlain by the Triassic Kidderminster Formation, and are provisionally assigned to the Permian.

Sedimentation during Triassic times was controlled by the east–west tensional stresses associated with continental rifting. In the west of the district, the deep Worcester Basin was an active depositional area in which thick sequences accumulated. The faults throwing down to the west and bounding the basin on its eastern margin, particularly the Woodgate, Stoke Pound and Lickey End faults (Figure 1), were active during sedimentation, so that thicker sequences occur on the downthrow sides than on the upthrow sides of these fractures. At this time, the Lickey Hills were, in contrast, an upstanding area across which most of the formations are thin. Farther to the east, the Knowle Basin extends as far as the Meriden Fault. Again, the Triassic rocks are thicker

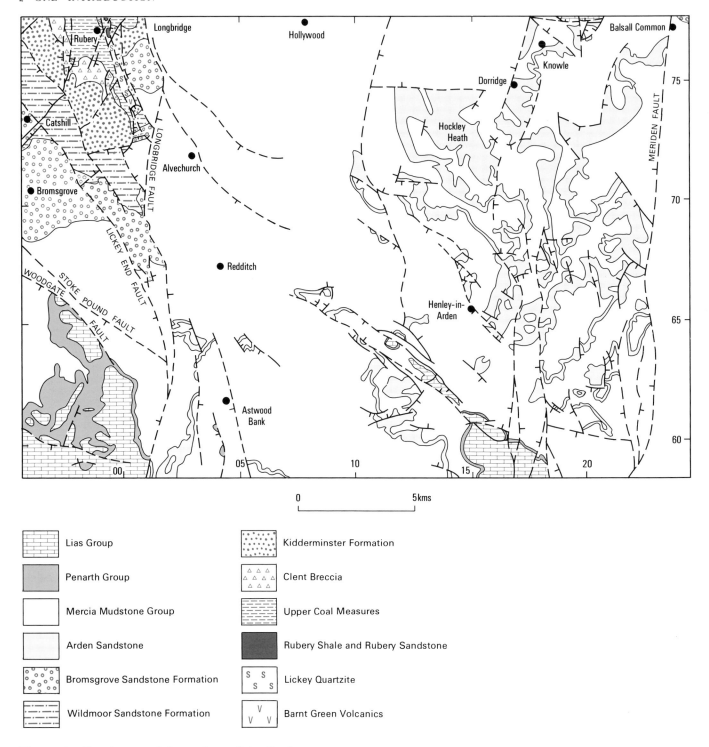

Figure 1 Sketch map of the geology of the district

here, but the deeper part of the basin has not been proved in detail. Geophysical evidence suggests, however, that it is not as deep as the Worcester Basin.

The Triassic strata of the district are divided into three major groups. The lowest, Sherwood Sandstone Group comprises three sandstone formations, the Kidderminster, Wildmoor and Bromsgrove Sandstone formations in ascending order. The Kidderminster Formation generally begins

with coarse basal conglomerates, especially well developed west of the Lickey Hills. They are largely the deposits of northerly flowing rivers, which introduced far-travelled, well-rounded pebbles. Just to the east of the Lickey Ridge, however, there are contemporaneous, locally derived breccias, which are scree deposits containing abundant blocks of Lickey Quartzite. The bulk of the Kidderminster Formation consists of coarse sandstone, deposited when the transporting power of the rivers decreased. The eastward limit of the formation is largely conjectural, although there is limited geophysical evidence that it might extend as far as the Meriden Fault. The Wildmoor Sandstone is red, fine grained and false bedded, with thin mudstone beds; it may be lacustrine in origin. It thins rapidly eastwards across the Lickey Hills and is overstepped by the Bromsgrove Sandstone at Longbridge. The thickness of the Bromsgrove Sandstone increases ten-fold westwards across the growth-faults bounding the Worcester Basin. The beds are largely sandstones, laid down by rivers flowing through an arid landscape or in ephemeral, shallow lakes. The uppermost beds were deposited by a brief marine transgression from the north-east and contain marine microplankton. These and other fossils show that the Bromsgrove Sandstone is of late Scythian to early Ladinian age.

The overlying Mercia Mudstone Group was deposited in an arid environment and is made up predominantly of blocky red mudstone, commonly with desiccation cracks and gypsum bands and nodules. The thickest accumulations were in the south-west. West of the Stoke Pound Fault a local evaporite basin developed in which the Droitwich Halite was laid down; this may extend into the present district at depth. The Arden Sandstone lies within the Mercia Mudstone sequence and marks a temporary change to fluvial conditions; it includes grey-green sandstone, siltstone and mudstone. Most of the fossils known from the group occur in the Arden Sandstone and are of Carnian age. Another amelioration of the arid climate is marked by the grey-green mudstone of the Blue Anchor Formation at the top of the succession.

The deposits of the overlying Penarth Group result from the major mid-Rhaetian marine transgression. The constituent formations are more varied in thickness and lithology than in other parts of the English Midlands because of contemporaneous earth movements. The Westbury Formation comprises black mudstone with variable proportions of pale grey sandstone. It contains a rich marine fauna, principally of bivalves. The overlying grey-green, calcareous mudstones of the Cotham Member lack a rich macrofauna but contain horizons with abundant marine microfossils.

Only the lowest portion of the Lias Group, represented by the Blue Lias Formation, occurs in the district. It consists of grey mudstones with sporadic limestones of marine origin. The basal part, below the lowest occurrence of *Psiloceras planorbis*, is placed in the Triassic, and the rest in the Hettangian Stage of the Jurassic. The uppermost preserved part of the sequence includes the Rugby Limestone in which thin limestones are abundant.

The absence of all younger solid formations from the district is due mainly to uplift and erosion during the Tertiary. The geological record is continued by the extensive Quaternary glacial drift of the Wolston Series and corresponding sediments. Drifts of eastern and western provenance are separated by a line through Balsall Common, Kingswood, Claverdon, Crabbs Cross and south along The Ridgeway. Postglacial erosion has removed considerable quantities of glacial drift and some of the debris is incorporated in the younger Quaternary deposits.

TWO

Cambrian and Ordovician

The Lower Palaeozoic rocks in the district include some of Tremadocian age, which are traditionally regarded by the British Geological Survey as being the youngest in the Cambrian System, and shown thus on the 1:50 000 map.

The oldest Lower Palaeozoic rocks include the Barnt Green Volcanic Formation and the Lickey Quartzite in the west, and the Merevale Shales in the east. The Barnt Green rocks were previously assumed to be of Precambrian age because of a supposed resemblance to other Precambrian sequences in the Midlands (Lapworth 1899, pp.328–330; Dunning in Harris et al., 1975, p.83). The discovery of burrowed mudstones and siltstones at Kendal End Farm [0034 7471] during the present survey instigated a search for microfossils at the same locality, and one sample yielded poorly preserved, broken and carbonised acritarchs referable to genera only. Dr S G Molyneux concludes that the predominance of *Acanthodiacrodium* and its association with *Polygonium* and *Stelliferidium* suggests a tentative Tremadoc age.

Although the stratigraphic relationship between the Barnt Green Volcanics and the Lickey Quartzite has not been observed, there is circumstantial evidence for an upwards-conformable passage from the former to the latter. Body fossils are unknown from the Lickey Quartzite, but it has in the past been correlated with the fossiliferous Hartshill Quartzite of proven Comley Series age (Lapworth, 1899, pp.350–351; Eastwood et al., 1925; Cowie et al., 1972). A thin mudstone bed exposed in a temporary section west of Rednal [9959 7650] yielded a sparse assemblage of poorly preserved acritarchs and scolecodonts. Dr S G Molyneux writes that Cambrian scolecodonts are unknown and the presence of *Baltisphaeridium?*, *Veryhachium* aff. *lairdii* and *V. trispinosum* s.l. suggests an Ordovician or younger age. The Lickey Quartzite is unconformably overlain by Llandoverian rocks, so that an Ordovician age is the most likely.

Correlatives of the Barnt Green Volcanics and Lickey Quartzite are unknown farther east, and the proving of Tremadocian mudstones at Meer End (see below) is in accord with the widespread occurrence of these rocks to the east (Old et al., 1987).

WESTERN AREA: LICKEY INLIER

Barnt Green Volcanic Formation

Water-laid crystal and crystal-lithic tuffs, together with tuffaceous sandstones, siltstones and mudstones, known collectively as Barnt Green Volcanic Formation, occur in a small, fault-bounded inlier extending southwards from Kendal End [003 747]. Apart from the outcrops at Kendal End, the formation is less well exposed now than when first surveyed in the last century, and little can be added to the field descriptions of Gibson and Watts (1898) and Lapworth (1899).

The beds are steeply dipping and generally strike north-west–south-east, roughly parallel to the major bounding faults of the Lickey Inlier; there is no consistent direction of dip, however, and many exposures are so weathered, shattered and veined that the bedding is difficult to distinguish in them. At Kendal End Farm, excavations have exposed beds dipping steeply to the east-south-east. The section shows about 25 m of coarse, dark green tuffs and tuffaceous sandstones, including 0.5 m of thin bedded, dark purple-grey, burrowed siltstones and mudstones intruded by a 30 cm-thick microdiorite sill.

Examination of thin sections largely confirms the descriptions given by W W Watts (Gibson and Watts, 1898); the significant differences are included below.

In the tuffs, the component mineral grains and rock fragments are predominantly angular, although plagioclase and quartz may form, respectively, euhedral and corroded phenocrysts; they have clearly not been transported far after ejection from the volcano. Plagioclase is nearly always the dominant feldspar, contrary to Watts' observation. The 'orthophyre tuffs' recorded by Watts appear to equate with the crystal-lithic tuffs; the latter closely resemble the crystal tuffs except that they also contain numerous rock fragments, including nonporphyritic andesite and altered porphyritic acid or intermediate lavas, some as pumice. The matrix of all these pyroclastic rocks consists predominantly of cryptocrystalline silica.

Sections of the laminated siltstone from Kendal End generally show an abundance of moderately well-sorted, angular quartz grains and muscovite flakes; plagioclase is present only in minor quantities. The few rounded quartz grains are probably derived phenocrysts, rather than water-worn fragments. Other grains include opaque minerals and carbonates, the latter possibly replacing plagioclase. The cement is a mixture of silica and opaque mineral, the latter sometimes abundant though largely absent from burrow-fills. Most of the tuffs and siltstones show extensive secondary veining by calcite or quartz.

The microdiorite intrusions are leucocratic, fine grained and generally nonporphyritic. They consist predominantly of plagioclase (oligoclase/andesine) in sheaf-like aggregates, commonly with a weak preferred orientation. Chlorite occurs interstitially and is presumably a secondary replacement of the rarely preserved hornblende. Opaque mineral is abundant in association with chlorite, and carbonate is common in veins and patches. One thin section shows a few slightly larger feldspar phenocrysts and more interstitial chlorite, giving an ophitic texture.

The 'brecciated porphyritic basalt' recorded by Gibson and Watts (1898) was not located during the present survey, and there are no hand specimens or thin sections matching this description among the local collections of the British Geological Survey.

Lickey Quartzite

The Lickey Quartzite crops out in the north-north-west-trending inlier of the Lickey Hills, in the north-west of the district between Kendal End [001 746] and Holly Hill [991 784]. It is a hard, brittle, jointed and very shattered rock, forming several low, steep-sided hills that are covered with a wash of quartzite chips and which support sparse vegetation. The inlier seems to be fault-bounded on all sides, except at Rubery where the Lickey Quartzite is overlain unconformably by the Rubery Sandstone or the Halesowen Formation. Elsewhere, its stratigraphical relationships are unclear and no confident estimate of its thickness can be given. Tuffaceous material occurs most commonly in what are probably the oldest beds exposed, and there may be an upwards passage from the Barnt Green Volcanics (Lapworth, 1899). The structure of the Lickey Hills is complex locally, with very variable dips, but in general an anticline trends parallel to the bounding faults of the inlier and plunges gently to the north-north-west. The steepest dips and overfolding occur mainly along the edges of the inlier and may relate to later movements along the bounding faults.

The petrography of the Lickey Quartzite has been studied by Watts (in Lapworth, 1899) and by Strong (1983). Strong's results are summarised in Table 1. The heavy minerals, notably glauconite, occurring in the Lickey Quartzite are listed by Fleet (1925, p.100).

There is no clear relation between the degree of sorting or the maturity of the sediments and their stratigraphic position in the Lickey Quartzite. The sorting, grain shape and sedimentary structures of the rock suggest deposition in a high-energy marine environment. Primary grain boundaries are still discernible and pressure welding is uncommon, suggesting early silica cementation. The presence of secondary chlorite, sericite and rare authigenic epidote indicates very low grade regional metamorphism.

Strata low in the sequence, exposed in a quarry [001 753] opposite Reservoir Road, Cofton Hill, comprise pale grey, brown and purple, flaggy, immature to submature quartzites in beds up to 0.6 m thick, interbedded with purple sand and micaceous shales. The colour of the quartzite is caused by finely disseminated, feldspathic, tuffaceous debris, and the shales are largely composed of the same material. This quarry exposes a synclinal overfold, with the beds folded about a near-horizontal axial plane (Plate 2) (Boulton, 1928, diagram p.256).

Ascending the sequence, the Lickey Quartzite becomes paler and incorporates less tuffaceous material. In the largest quarry, in Rednal Gorge [998 759], massive beds of dark purplish grey quartzite, each up to 1 m thick, are separated by yellowish green and deep purple, sandy clay partings. At the disused Leach Green quarry [995 769] and at the Bristol Road south cutting [992 774], the Lickey Quartzite varies from fine grained and white, to coarse, grey and pebbly, and is in massive beds up to about 1 m thick, which were lithified and jointed before the transgression of the Llandovery sea, because sands of Llandovery age have infiltrated down cracks. The formation here is cut by a very weathered dyke which is truncated by, and thus older than Llandovery strata.

EASTERN AREA: WARWICKSHIRE COALFIELD

Merevale Shales

The Merevale Shales have been proved within the district only in the Meer End Borehole [2406 7447] where they were penetrated for 14 m to the base of the hole at 1043 m. They comprise moderately dipping, grey, bioturbated silty mudstone with thin sandstone and siltstone bands, and lithologically resemble the Merevale Shales of the type area

Table 1 Petrography of the Lickey Quartzite

	Composition	Sorting	Specimen	Locality
IMMATURE Lithic arenites	Q/F/RF/MP/C	p	E58477	Cofton Hackett
			E58478	Cofton Hackett
	Q/F/RF/C	p-ms	E2939	Rubery
ARKOSIC Mainly subarkoses 10-25% F	Q/F	p-ms	E58474	Reservoir Road
			E58475	Reservoir Road
			E58482	Lickey
SUBMATURE Feldspathic quartzites 10% F	Q/F	p-ms	E58476	Cofton Hackett
			E58476	Lickey
			E2938	Keepers Lodge
			E11573	Rubery Hill
Quartzites	Q/C	p-ms	E58480	Lickey
			E58485	Rubery Hill
MATURE Mature quartzites	Q	ws	E58479	Kendal End
			E58483	Lickey
			E58484	Leach Green

Q – quartz, F – feldspar, RF – rock fragments, MP – chloritic mud pellets, C – cherts, p – poorly sorted, ms – moderately sorted, ws – well sorted

Plate 2 Lickey Quartzite in south-west face of a quarry at Cofton Hill [001 753] viewed in 1921.

Massive quartzite beds with thin bands of purple mudstone, folded into a synclinal reclined fold with its axial plane dipping out of the picure to the left. (A2022)

near Nuneaton (Taylor and Rushton, 1971). Fossils in the mudstones have been identified by Dr A W A Rushton and include trace fossils, *Rhabdinopora flabelliformis flabelliformis*, *R. flabelliformis patula*, *Eurytreta sabrinae* and *Lingulella* sp., indicating a Tremadoc age.

THREE

Silurian

Strata of Silurian age are present in small inliers and at depth in the north-west of the district. They range from late Llandovery to upper Wenlock in age. By comparison with the Llandovery and Wenlock sections of the South Staffordshire Coalfield (Butler, 1937; Hamblin et al., 1978) they are probably part of a conformable sequence about 330 m thick.

The lowest formations in the succession are the Rubery Sandstone and the overlying Rubery Shale, both late Llandovery in age. Both formations contain a marine fauna including *Stricklandia laevis*, indicative of the Telychian Stage. The Rubery Shale also contains brachiopods, corals, trilobites and graptolites including *Monograptus marri* and *Pristograptus nudus* (Wills et al., 1925).

Rubery Sandstone

The Rubery Sandstone overlies the Lickey Quartzite unconformably, and crops out at Rubery Hill Hospital [993 778] and Rubery [993 773]; the presence of a third inlier at Roseleigh Road [999 767], recorded by the previous survey, has not been confirmed. Exposures in a quarry [9927 7727] south of Bristol Road show massive, coarse-grained, decalcified sandstone, varying from pale grey and compact to open textured, with reddish and purple stains. Some of the constituent grains are well rounded and probably aeolian in origin. The basal bed contains clasts of Lickey Quartzite up to 15 cm across, and locally fills hollows in the irregular surface of the underlying quartzite. Sand-filled 'neptunean dykes' up to 20 cm across extend down into fissures in the quartzite. The beds were deposited in a sea transgressing a rugged, arid shoreline. They probably buried the Lickey Quartzite ridge completely, but Rubery Sandstone fragments occurring in the basal conglomerate of the Halesowen Formation in Leach Green Lane [9939 7720] (Boulton, 1928) afford the only evidence of Silurian strata west of the ridge.

The full sequence at Rubery was exposed when Bristol Road was widened (Wills et al., 1925). It is 31.7 m thick, the upper part including white, red and purple shales interbedded with fine and coarse sandstones. North of the mapped outcrops, the Rubery Sandstone continues at shallow depth beneath rocks of the Halesowen Formation. It was previously exposed, resting upon Lickey Quartzite, at Hollyhill Quarry [c.9910 7838] (Eastwood et al., 1925, p.13), and fragments of red sandstone, which are probably Rubery Sandstone, have been recorded at Kendal End (Lapworth, 1899). Heavy minerals occurring in the Rubery Sandstone are listed by Fleet (1925, p.104).

Rubery Shale

The Rubery Shale succeeds the Rubery Sandstone conformably in inliers at Rubery and Rubery Hill Hospital, where it is overlain unconformably by the Halesowen Formation. Wills et al., (1925) describe the lowest 23 m of strata along Bristol Road as buff, grey, blue and purple, noncalcareous shales (with 'fucoids' at the base), interbedded with thin beds of fine-grained, decalcified, fossiliferous sandstone and thin, white and purple limestone. Later excavations are described by Wills and Laurie (1938). Strata higher in the sequence, formerly regarded as of Wenlock age (Eastwood et al., 1925) but now recognised as of late Llandovery age (Ziegler et al., 1968, p.764), are visible in Callow Brook [9929 7762]. They are pale grey, buff and purple shales with beds of hard, fine-grained, pale grey, crystalline limestone up to 15 cm thick.

Wenlock Shale

The only strata of early Wenlock age have been proved buried beneath the Halesowen Formation east of Rubery. A layer of impure limestone 0.3 m thick, found in a sinking for coal at 'Colmers' [c.997 775] (Murchison, 1839, p.493; Yates 1829. p.258), is possibly part of the local correlative of the Woolhope Limestone of the Welsh Borders or the Barr Limestone of Staffordshire, both of early Wenlock age. Silurian mudstones ('blue binds') proved for 42.4 m at Longbridge Pumping Station [0072 7755] have also been assigned to the Wenlock Shale (Boulton, 1928); probably the total thickness is similar to the 272 m proved at Walsall (Butler, 1937).

Wenlock Limestone

Rocks of late Wenlock age (Bassett, 1974, p.754) crop out at Kendal End [003 746] in a triangular, fault-bounded inlier, and are assigned here to the Wenlock Limestone. Murchison (1839, p.493) records a steeply dipping limestone here, less than 1 m thick, and also shales with limestone nodules, in quarries then recently abandoned. The quarries had been completely infilled by the time of the present survey in 1983. Hardie (1954) delineated the outcrop by augering and dug two trenches, one of which revealed greyish shales with a 0.1 m bed of nodular, calcite-veined limestone. He also noted (1954, p.149) an exposure of limestone [0011 7464] which is interpreted here as a thin slice of strata caught in a fault between the Barnt Green Volcanic Formation and the Keele Formation.

FOUR

Carboniferous

There are marked differences between the Carboniferous and Permian sequences in the east and west of the district (Table 2). The eastern part of the district lies on the south-western flank of the Warwickshire Coalfield where an apparently conformable sequence of about 940 m of Coal Measures rests unconformably on the Merevale Shales. The north-western part of the district lies at the southern end of the South Staffordshire Coalfield, but the productive Coal Measures are absent, having been overstepped by about 400 m of Upper Coal Measures. The successions in the two coalfields are shown on separate vertical sections on the 1:50 000 map.

Lower and Middle Coal Measures, and Etruria Marl

These rocks are absent in the west, and even in the east have been proved only in the Meer End Borehole [2406 7447] (Figure 2). The sequence in the borehole was heavily faulted and the conjectural thicknesses shown on the map sections are based on provings in the adjacent districts to the north and east.

The Lower and Middle Coal Measures are predominantly grey mudstone and siltstone with minor seatearths and one particularly important coal, the Thick Coal. This seam is a composite of a number of seams which are more widely separated and individually named farther north (Cope and Jones, 1970). None of the marine bands that are used to subdivide the succession elsewhere were proved in the Meer End Borehole.

North of Coventry the Vanderbeckei (Seven Feet) Marine Band occurs only a few metres below the Thick Coal (Mitchell, 1942), and it is, therefore, assumed that the lowest 10 m of Coal Measures at Meer End belong to the Lower Coal Meaures. It is likely that the Aegiranum (Nuneaton) Marine Band occurs in the sequence since it is present farther east, but it was not proved in the borehole.

The Etruria Marl has a gradational basal contact with the productive Coal Measures. It consists of mudstone with thin siltstone beds and, at the top, a 1.5 m pebbly 'Espley' sandstone, containing abundant rounded and angular clasts of mudstone, siltstone and ironstone. Shades of grey, red and brown are general, and may be combined in some beds to produce a variegated appearance.

Halesowen Formation

In the north-west of the district, the Halesowen Formation is about 30 m thick. It is largely made up of micaceous generally pale grey, but locally mottled red, yellow and purple mudstones, and greenish grey, micaceous sandstones. At the base there is up to 0.9 m of a locally reddened conglomerate carrying angular blocks of Lickey Quartzite and Rubery Sandstone in a clay matrix. There is at least one limestone bed of *Spirorbis*-type, although no fossils have been recorded, and at least one coal seam, 0.3 to 0.6 m thick but splitting locally into interbedded coals and shales. The base of the Halesowen Formation probably approximates to the base of the Westphalian D Stage (Ramsbottom et al., 1978).

The formation crops out on either side of the north end of the Lickey Hills, where it rests unconformably upon a buried topography of Lower Palaeozoic rocks, and apparently thins

Table 2 Comparative Westphalian stratigraphy (based on Ramsbottom et al., 1978 and Old et al., 1987)

	Lickey Hills	Warwickshire Coalfield	
PERMIAN	Clent Breccia	Ashow Formation and Kenilworth Sandstone Formation (sheet 184)	Enville Group
CARBON-IFEROUS Westphalian D	unconformity	Tile Hill Mudstone Formation	
	Bowhills Formation	Coventry Sandstone Formation	
	Keele Formation	Keele Formation	
	Halesowen Formation	Halesowen Formation	
Westphalian C	unconformity	Etruria Marl Formation	
Westphalian B		Middle Coal Measures	
Westphalian A		Lower Coal Measures unconformity	

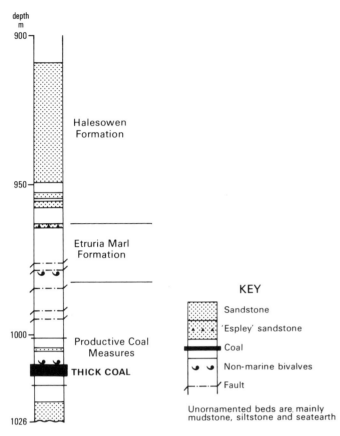

Figure 2 Coal Measures succession cored in the Meer End Borehole

as it transgresses onto them. Dips are gentle, except for local steepening against faults. Coal has been widely recorded at outcrop or in shallow borings and excavations, though probably only a single seam is present. Diverse, well-preserved miospores of Westphalian D age have been identified by Dr B Owens from this coal and its shale roof at outcrop in Callow Brook [9967 7768]. A complete list of the miospores is given in the open-file report for SO97NE; the following association of species, particularly *Thymospora* spp., indicates a Westphalian D age: *Cadiospora magna, Microreticulatisporites nobilis, Punctatosporites granifer, P. oculus, Spinosporites spinosus, Thymospora obscura, T. pseudothiesseni, T. thiesseni, Torispora securis, Triquitrites sculptilis* and *Vestispora fenestrata*. Gibson (unpublished manuscript) recorded a 15 cm coal smut that crops out near the road from Colmers Farm [000 775] to Frogmill Farm [997 787], and which lies 'a few feet' above a 3 to 5 cm bed of nodular limestone.

At the south end of the Lickey Hills, the Halesowen Formation is represented by a small outcrop of blue-grey, silty clays at Kendal End Farm [0011 7464], recorded by Hardie (1954, p.14), which are faulted against Wenlock strata and pass up into the Keele Formation, which it probably everywhere underlies. A borehole at Longbridge Pumping Station [0072 7755] records 17.6 m of Halesowen beds resting upon Wenlock Shale, and apparently faulted against the overlying Kidderminster Formation; the coal known within the sequence throughout the area may be recorded as 'blue and black marl' 16.1 m above the base. In the east, the lower

63 m of the Halesowen Formation cored in the Meer End Borehole (Figure 2) comprise grey sandstone, mudstone and siltstone, with a few thin coals. Coaly stems and other plant debris are commonplace. The sandstones are typically fine to medium grained and massive, in beds up to 16 m thick. They include a few pebbly bands with clasts of ironstone, siltstone and mudstone. A few show cross-bedding and siltstone laminae. The siltstones are well laminated, and some exhibit ripple drift cross-lamination and slump structures. The mudstones are mainly greenish grey and are commonly highly slickensided.

Keele Formation

These rocks succeed the Halesowen Formation conformably, and comprise dominantly red, orange and variegated marls (calcareous mudstones) with large, green reduction spots ('fisheyes'), and thin red, brown and purple sandstones. The sandstones commonly contain penecontemporaneous mudstone conglomerates, but no extraformational clasts. At least one *Spirorbis* limestone occurs.

The formation occupies low ground to the west of the Lickey Hills, but the measures thin rapidly southwards beneath the unconformable Clent Breccias and Kidderminster Formation, and are overstepped at Barnt Green [0035 7315]. Exposures of the soft marls are few, but a thin sandstone crops out across the golf course at Rubery [9903 7639 to 9933 7571]. *Spirorbis* limestone outcrops were recorded beside the Whetty Brick and Tile works [9802 7713] and on the golf course [9927 7646]; their stratigraphical relationship with the three *Spirorbis* Limestone beds which occur in the Birmingham area (Eastwood et al., 1925) is not known.

East of the Lickey Hills, along Bristol Road [995 775], Wills and Laurie (1938) detailed about 60 m of Keele mudstones with sandstone beds overlain unconformably by the Kidderminster Formation, and farther along the same outcrop a *Spirorbis* limestone is known low in the sequence (Yates, 1829, p.256; Boulton, 1925). The continuous Keele outcrop along the south-east flank of the South Staffordshire Coalfield enters the district on the east side of the Frogmill Fault [9988 7780], and the formation probably continues at depth across the district to the Warwickshire Coalfield.

In the east of the area, the limits of the Keele Formation have been determined in the Meer End Borehole from the downhole geophysical logs. Chipping samples indicate a red-brown, predominantly mudstone sequence with interbedded sandstones.

Enville Group—Bowhills, Coventry Sandstone and Tile Hill Mudstone formations

In the west of the district, the Keele Formation passes conformably up into the Bowhills Formation, which is in turn overlain by the Clent Breccias of presumed Permian age. The Bowhills Formation (Ramsbottom et al., 1978, pl. 3), previously termed the Enville Beds or Calcareous Conglomerates, is thought to be of Westphalian D age, although palaeontological evidence is absent from the present area and equivocal elsewhere, and the formation may be at least partly Permian. It comprises thick, massive, red and green sandstones and red mudstones, the former locally containing con-

Plate 3 Conglomerates and pebbly sandstones in the basal conglomerate of the Kidderminster Formation, Marlbrook Quarry [982 747].

The largest boulders come from the very base of the formation and rest on Clent Breccias, which is visible as a disturbed clay at the bottom left. (A13503)

glomerates with clasts of Dinantian limestone, quartzite, vein-quartz and pyroclastic rocks (King, 1899, p.118), and with a calcareous cement.

The Bowhills Formation forms a faulted outcrop between Windmill Hill [974 779] and Eachway [988 766]. Between Eachway and Beacon Wood [1979 759], the lowest 45 m of the formation include four sandstones and are preserved in a block which was down-faulted before the deposition of the Clent Breccias. Disused quarries [9838 7667] in Beacon Wood expose medium-grained, soft, massive, non-micaceous sandstones; the highest sandstone exposed includes clasts of chert up to 2 cm across. The outcrop west of the Holywell Fault comprises mainly mudstones lying higher in the sequence, but includes a 0.8 m exposure of red, medium-grained, massive, micaceous sandstone at Waseley-hill Farm [9754 7752]. It is possible that the basal sandstone of the formation crops out above the Keele Formation in the railway cutting at Barnt Green [0035 7320], where Strickland (1842) described a 64 m-long exposure of brown, red and grey, feldspathic sandstone with rolled fragments of dark red, indurated marl.

The Coventry Sandstone Formation is the correlative in the east of the Bowhills Formation, and similarly overlies the Keele Formation conformably. In the Meer End Borehole it has been delimited by geophysical logging. The succession differs from the Keele sequence in comprising mainly thick sandstones with subordinate beds of mudstone. The Tile Hill Mudstone Formation succeeds conformably and has a tiny outcrop in the north-eastern corner of the district. This is contiguous to the east and north with the more extensive outcrops of a 300 m-thick mudstone sequence with thin sandstone beds. The same lithologies have been identified from the geophysical logs and chipping samples of the Meer End Borehole. Red-brown, sandy mudstones, also tentatively assigned to the Tile Hill Mudstone, were penetrated beneath the Bromsgrove Sandstone in boreholes at Shrewley [222 6755] and Heath End [233 595]. Detailed accounts of the Enville Group of the Warwickshire Coalfield are given by Eastwood et al. (1923) and by Old et al. (1987).

FIVE

Permian

Clent Breccias

The Clent Breccias represent the upper part of the Enville Group in the western part of the district, where they rest unconformably on the Bowhills and Keele formations. They are composed of red-brown and purple proximal breccias and fanglomerates, with generally angular clasts of sandstone, shale, basalts, rhyolites, tuffs and Carboniferous dolerites, all in a clay matrix. Secondary calcite veins are common. The clasts at Beacon Hill [989 769] are entirely volcanic and apparently derived either from outcrops which are no longer exposed, or from the Barnt Green Volcanic Formation (Boulton, 1933). They reach 0.6 m across but are generally between 0.13 and 0.23 m in diameter. Farther north, a few fragments of Lickey Quartzite, up to 0.38 m across, were noted. Heavy minerals occurring in the Clent Breccias were listed by Fleet (1925, p.119). The breccias coarsen towards the south-south-east, and represent coalescing fans derived from mountainous terrain only a short distance away in this direction. The subsidiary 'Worcester Horst' (Wills, 1956, p.101) may locally have contributed clasts from the west.

The Clent Breccias reach a maximum thickness of about 150 m in the north-west of the district, but are progressively cut out to the south and east beneath the unconformable Kidderminster Formation until they are completely overstepped in Pinfields Wood [997 743]. Although the formation forms the highest ground in the area, at Beacon Hill and Waseley Hill, exposures of the breccias break down rapidly as the soft clay matrix separates from the megaclasts. Deep purple breccia is exposed in places beneath the basal conglomerate of the Kidderminster Formation at Marlbrook Quarry [982 747], its upper surface reduced to very pale green by water percolating through from above (Wills, 1945). Nearby, boreholes for the M42 motorway proved up to 17.45 m of breccia, while in Burcot No. 3 borehole [9848 7173], breccias dipping at 35° were proved from 190 m to 244 m depth. The formation has not been proved south of Burcot or south-east of its present outcrop.

SIX

Triassic

The greater part of the district is underlain by Triassic strata belonging to the Sherwood Sandstone, Mercia Mudstone and Penarth groups. Deposition took place under generally arid and semiarid conditions, in a low-latitude continental interior. Most of the strata are red as a result of the diagenetic alteration to iron oxide (haematite) of detrital ferromagnesian silicates and iron-bearing clay minerals (Walker, 1976).

SHERWOOD SANDSTONE GROUP

This name was formally introduced (Warrington et al., 1980) for the formations that comprise the arenaceous lower part of the Triassic succession throughout Britain. This sequence was subdivided by Hull (1869) into three formations, renamed recently (Warrington et al., 1980), which are the basis for the classification used in this account. An additional formation, the Quartzite Breccia, which locally underlies the Kidderminster Formation, is here included in the group.

SUBDIVISIONS OF THE SHERWOOD SANDSTONE GROUP

Hull, 1869	*This account*
Lower Keuper Sandstone	Bromsgrove Sandstone
Upper Mottled Sandstone	Wildmoor Sandstone
Bunter Pebble Beds	Kidderminster Formation
	Quartzite Breccia

Deposition of the Sherwood Sandstone Group was controlled by palaeogeographical changes initiated during the Permian (Audley-Charles, 1970). A series of asymmetric troughs and ridges, orientated roughly north–south and bounded by listric faults, formed by continental rifting in response to east–west tensional stresses in the region of the present North Atlantic. One such trough was the Worcester Basin (Wills, 1956; Chadwick and Smith, 1988), the northeastern flank of which lies on the western side of the district. Another, the Knowle Basin (Wills, 1956), which extends from the Longbridge Fault to the Warwickshire Coalfield, appears to be shallower than the Worcester Basin, but the full depth has not been proved. The Lickey End and Longbridge faults (Figure 3) and the intervening parallel structures are believed to have been active in early Triassic times, so defining the edges of these depositional basins.

The downwarping of the Worcester Basin resulted in a river system bringing detritus from as far south as Brittany. The lower fluvial part of the Kidderminster Formation is restricted to the Basin area, while the Quartzite Breccia formed as a scree deposit on the eastern flank of the ridge (the 'Lickey Axis' of Wills, 1948) between the Worcester and Knowle basins. However, this ridge was soon inundated by the upper part of the Kidderminster Formation, and although it subsided more slowly than the flanking basins during the remainder of Sherwood Sandstone deposition, it

did not then constitute a barrier to sedimentation. The Sherwood Sandstone Group is about 700 m thick in the west of the district; in the east the thickness is less well known, but there is geophysical evidence for an eastward thinning, across the Meriden Fault, from about 600 m to about 60 m (Allsop, 1981).

Quartzite Breccia

The Quartzite Breccia is up to 10 m thick and is composed entirely of angular blocks of Lickey Quartzite, up to 0.3 m in diameter and of very local origin, with little sand or clay matrix. Wills and Shotton (1938) discovered the breccia in a trench in Tessal Lane [006 785] (in the adjoining Birmingham district), where it rests upon Keele Formation. It has a very limited outcrop (Figure 3) and is currently not well exposed.

The Quartzite Breccia passes laterally into sandstones in the lower part of the Kidderminster Formation, in which there are interbedded thin seams of breccia. The latter thin to the north-west and point to a nearby source to the southeast, possibly an outlier of Lickey Quartzite along the line of the Longbridge Fault. The Quartzite Breccia is envisaged as a scree to the east of the Barnt Green Fault, contemporaneous with the basal conglomerate of the Kidderminster Formation in the adjacent Worcester Basin[1].

Smith et al. (1974, table V, column 24) equate the Quartzite Breccia with the Bridgnorth Sandstone of early Permian age. The present authors consider it to be Triassic and a part of the Sherwood Sandstone Group, as it passes by intercalation up into the Kidderminster Formation. No fossils are known from the formation.

Kidderminster Formation

This name was introduced (Warrington et al., 1980) for beds formerly termed Bunter Pebble Beds. In this district the sequence of fluvial sandstones and conglomerates crops out in the north-west, where it attains a thickness of about 175 m; it is present at depth in the south-west and north-east, where it may reach a thickness of 375 m, but is absent in the south-east. In the west of the district it comprises a basal conglomerate overlain by a mainly sandstone sequence, but east of the Longbridge and Rednal faults the conglomerate is absent.

SEDIMENTOLOGY AND MODE OF DEPOSITION

The Kidderminster Formation comprises a sequence of upward-fining rhythms, each with an erosional base. The

---

1 The representation of the Quartzite Breccia as overlying the basal conglomerate of the Kidderminster Formation on the 1:50 000 map is misleading.

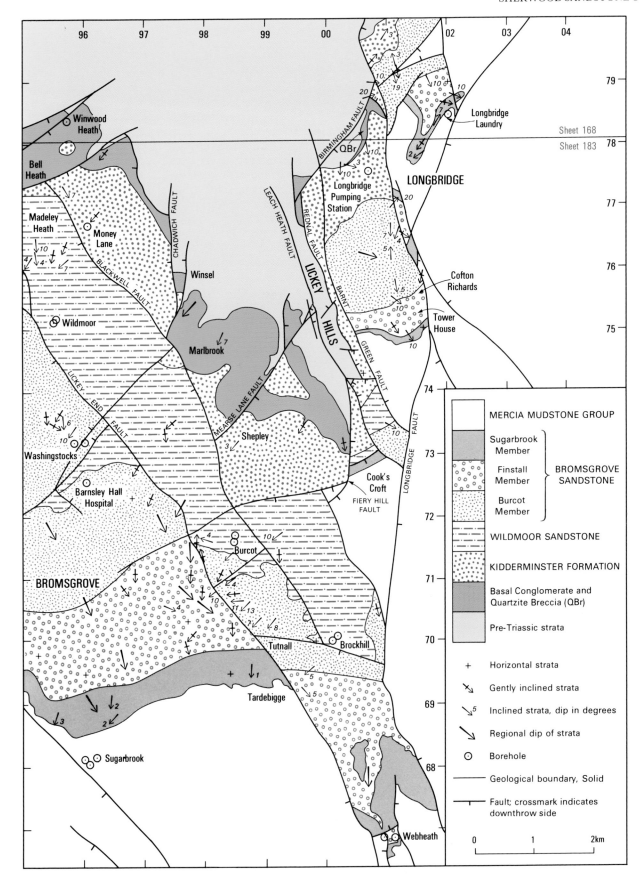

Figure 3 Outcrops of the formations and members of the Sherwood Sandstone Group

rhythms commence with a hard conglomerate, overlain by sandstone which is succeeded by mudstone; they are rarely complete, the mudstone commonly being absent. The average grain size of the sediments decreases upwards through the formation. The basal beds are mainly conglomerate with only subordinate sandstones, while the succeeding beds are dominantly sandstones with subordinate conglomerates and mudstones.

The basal conglomerate is composed of pebbles and cobbles in a weakly cemented matrix of coarse, micaceous, moderately well-rounded sand (Plates 3 and 4). Approximately 65 per cent of the pebbles are quartzite and 25 per cent vein quartz; other rocks constitute 10 per cent. The quartzites include white to dark grey and dull red or 'liver coloured' varieties. Many are locally derived Lickey Quartzite, moderately to poorly rounded and up to 57 cm across.

The remainder are smaller, well rounded and often pressure-pitted (the 'Bunter' quartzite of earlier literature). Lamont (1946) lists Ordovician, Silurian and Devonian faunas obtained from pebbles in the Kidderminster Formation at Birmingham; noteworthy amongst these fossils is the Ordovician *Orthis budleighensis* (now *Tafilaltia valpyana*). Wills (1945, p.24) recorded *Spirifer* (now *Cyrtospirifer*) *verneuili*. Lamont concluded that the Ordovician and Devonian pebbles were derived either from Brittany or Cornwall and that the Silurian pebbles originated from the Upper Llandovery of the west Midlands.

It is now generally believed (see discussion in Fitch et al. 1966, p.301; Holloway et al., 1989) that the well-rounded pebbles were introduced by a major river flowing from the Hercynian Highlands to the south of the district (the 'Budleighensis' River of Wills, 1956), joined by tributaries

Plate 4 Close-up view of conglomerate in Kidderminster Formation, Marlbrook Quarry [982 747].

Most of the clasts are well-rounded 'Bunter' quartz and quartzite pebbles and cobbles. The jointed pebble to the right of the hammer is a Dinantian chert. In each bed the pebbles coarsen downwards, the lower beds here being not far above the Clent Breccias. (A13504)

rising in upland areas to the south-east and south-west. Even Wills' suggestions that they may derive from a pre-existing conglomerate (1956, p.113) or from a concealed exotic klippe (1970a) would not alter the picture of a major braided river flowing northward through the Worcester Basin (Fitch et al., 1966, p.305) to discharge into widespread alluvial fans in Shropshire, Staffordshire and Cheshire. At the time of formation of the basal conglomerate, this river was restricted to the area west of the Longbridge and Rednal faults, but was fed by tributary streams from the west and east that contributed locally derived material to its load of exotic pebbles. The course of the river was presumably controlled by contemporary growth faulting (Whittaker, 1985; Chadwick and Smith, 1988). The high proportion of local Lickey Quartzite in the conglomerate may mean that the Lickey Hills formed an upstanding area for at least part of the period of conglomerate deposition, or that these local quartzites were derived from similar faulted inliers farther south that are now buried beneath the Trias. After the formation of the basal conglomerate, the river system rapidly overstepped its faulted margin eastward and covered a much wider area, the increased width of the braided system and consequent decrease in transporting power probably being the reason for the decrease in grain size.

The other pebbles in the conglomerate are a mixture of exotic and local lithologies including quartz-tourmaline rocks ('schorl'), Dinantian limestones and cherts (up to 23 cm long recorded), grits and greywackes (to 22 cm), claystones (to 20 cm), Llandovery sandstone, soft purple porphyries and hard angular volcanic rocks (to 10 cm), the last possibly basalt from the Clent Breccias. Schorl and certain quartz porphyries can be matched with rocks in Devon and Cornwall (Smith, 1963). At Wildmoor, a 20 cm diameter armoured clay ball was recorded, and the basal conglomerate includes beds, up to 0.9 m thick, of coarse, cross-bedded sandstone.

The main part of the Kidderminster Formation is dominated by massive red-brown to yellow-brown sandstones with a weak calcareous cement. The sand grains are coarse to fine in grade and largely subangular, but with some 'millet seed' grains. The sandstones are moderately well sorted and micaceous, and contain large amounts of weathered feldspar; heavy minerals are listed by Fleet (1923). They are largely cross-bedded, though some planar-bedded units occur near the tops of rhythms. Other structures include slumps, mudcracks and ripple marks. Conglomerates of similar lithology to the basal conglomerate occur at the base of many of the rhythms, but upwards through the sequence they decrease in thickness, frequency and pebble size. Scattered pebbles comprising a suite similar to that in the basal conglomerate are also present, but are smaller and less well rounded, and include a higher proportion of local rock types; quartzites reach 15 cm in length and other lithologies 7 cm. Mudstone beds up to 0.8 m thick occur at the top of some rhythms, but are uncommon. They are brick-red and may be either blocky or bedded and sandy, with very micaceous bedding planes. Mudstone clasts up to 20 cm across are common in the conglomerates and sandstones and probably represent penecontemporaneous erosion of such mudstone beds.

The sedimentary structures and petrology of the fining-upward cycles of the Kidderminster Formation support the view that it is fluvial (Allen, 1965; Pettijohn et al., 1972, p.456). The conglomerates represent channel lag deposits and gravel bars, the massive cross-bedded sandstones formed in large migrating sand-bodies within fluvial channels, and the planar-bedded sandstones and mudstones represent levée and overbank deposits. The dominance of channel deposits suggests a large number of low-sinuosity, braided stream courses rather than a more mature meandering system. The coarseness and thickness of the cycles, and the degree of crevassing (basal-bed downcutting), suggest considerable variation in flow rate, with all the channels being occupied in times of flood and most being abandoned in times of drought (Collinson, 1986). The change in facies from the poorly cross-bedded gravels of the basal conglomerate to the strongly cross-bedded but still poorly cyclic sandstones of the bulk of the formation may be interpreted using Miall's (1977) classification of braided streams. The former are believed to represent the Scott type of braided stream, which is of proximal gravelly rivers, while the latter represent the Platte type, laid down in very shallow rivers or those without marked topographic differentiation. The change may be occasioned by a decrease in angle of fall of the river, or a decrease in precipitation in the source area.

A semi-arid climate with alternating wet and dry periods is implied for the uplands to the south, but an arid climate possibly prevailed in the West Midlands (Wills, 1956, p,112; Audley-Charles, 1970, p.68). The presence of millet seed grains in the sandstones suggests the presence of desert sand dunes in the area, but no sandstones of aeolian origin are recorded and it is assumed that these grains were blown into the river, or were reworked from older deposits.

PALAEONTOLOGY

No fossils have been found in the Kidderminster Formation within the district. However, trace fossils are known from the formation in boreholes at Bellington [877 768], some 8 km to the west, in the adjoining Droitwich (geological sheet 182) district (Wills, 1970b; Wills and Sarjeant, 1970). These comprise invertebrate traces (*Permichnium völkeri*) and vertebrate tracks assigned to the ichnogenera *?Aetosauripus* sp., *Coelurosaurichnus* spp., *Hamatopus* sp., *?Procolophonipus* sp., and *Rhynchosauroides* sp.. Demathieu and Haubold (1972) considered the vertebrate tracks to be referable only to the ichnogenera *Procolophonichnium* sp. and *Rhynchosauroides* sp., but the assemblage nevertheless provides unique evidence of the existence of a diverse land fauna, including insects and reptiles, in the region during deposition of the Kidderminster Formation. The reptile association is believed to have included pseudosuchians, coelurosaurs, rhynchocephalians and possibly cotylosaurs, and is compatible with environmental interpretations (see above) based upon lithological and sedimentological evidence.

STRATIGRAPHY

The Kidderminster Formation rests unconformably on the Enville Group, Keele and Halesowen formations and

Rubery Sandstone, and must have overstepped onto Lickey Quartzite; locally it rests conformably on the Quartzite Breccia. Wills (1970a, 1976) subdivided the Kidderminster Formation and Wildmoor Sandstone into five 'miocyclothems' of which the basal conglomerate is the lower part of the lowest ('BSI'). Each is subdivided into microcyclothems which correspond to the rhythms described here.

The formation was proved at depth in the west in a series of boreholes for water at Wildmoor, Washingstocks, Barnsley Hall Hospital, Burcot, Brockhill and Webheath (Figures 3 and 4). The base was penetrated only by Burcot No. 3 Borehole, which proved a thickness of 10.8 m of basal conglomerate with pebbles up to 25 cm across. However, Brockhill No.2 Borehole proved 21.6 m of basal conglomerate without penetrating its base. The Wildmoor boreholes yielded artesian flow from these conglomerates, demonstrating that there are impermeable strata higher in the formation. The strata overlying the basal conglomerate are composed of upward-fining rhythms, dominated by coarse- to fine-grained, cross-bedded sandstones, with basal conglomerates. These strata are thickest (155 m) at Wildmoor, and thin to 133 m at Burcot and 127–129 m at Brockhill. There is some suggestion that the upward-fining rhythms can be grouped into larger rhythmic units, and Wills (1970a, 1976) correlated five 'miocyclothems' between the boreholes. He originally distinguished four such units (BSI to IV) but subsequently considered that his BSIIA and BSIIB divisions of BSII are cycles in their own right (1970a, p.235; 1976, p.49). Ripple marks and desiccation cracks in mudstone laminae are recorded at Wildmoor (Wills, 1970a, p.236), and slump structures at three levels at Webheath (Figure 4).

The Kidderminster Formation in the north-west of the area has a faulted 'U'-shaped outcrop, from Winwood Heath around the southern end of the Lickey Hills to Longbridge, in the form of an anticline plunging gently south-south-east (Figure 3). Winwood Heath Borehole [9568 7839] (in the adjoining Birmingham district) penetrated about 38 m of basal conglomerate without proving its base. South-eastwards, the conglomerate thins to 17 m at The Winsel [975 755] and 16 m at Old Birmingham Road [981 743]. At Marlbrook Quarry [982 747], a 9 m face resting on Clent Breccias was composed of beds of cross-stratified gravel up to 3 m thick, alternating with beds of sand up to 1.1 m thick, with the largest pebbles (of Lickey Quartzite up to 57 cm across) restricted to the lowest bed. East of the Mearse Lane Fault, the basal conglomerate continues to thin, but it is not known whether it extends east of the Lickey Hills at Barnt Green, as the base of the Kidderminster Formation is faulted out east of the Fiery Hill Fault. The reappearance of the conglomerate east of the Rednal Fault demonstrates that at least the northern end of the Lickey Hills was buried early in the Triassic. Wills and Laurie (1938, p.176 and pl. I) recorded 3.7 m of 'typical rounded and pressure-pitted pebbles with only a little sand' beside Bristol Road [9956 7750]. However, 110 m farther east, at Rubery [9967 7749], they recorded sandstone with only scattered pebbles resting on the Keele Formation; the basal conglomerate is absent east of Rubery.

The sandstone-dominated upper part of the Kidderminster Formation has been worked for building sand and is exposed in quarries at Upper Madeley Farm [957 731]. The working face at Shepley [9844 7310] reveals 25 m of remarkably uniform, brownish red, cross-bedded, medium- to coarse-grained, micaceous and feldspathic sandstone, with a few lenticular beds of blocky red mudstone up to 15 cm thick. Scattered clasts of red mudstone occur in the sandstones, together with a few pebbles of quartz and quartzite up to 3 cm in size and loosely grouped in bands. Boreholes prove the thickness of almost pebble-free sandstones here to be at least 41 m, while boreholes at Money Lane [960 767] prove up to 60.5 m of similar sandstone with a few mudstone beds up to 0.5 m thick, scattered pebbles, and one conglomerate bed about a metre thick. The lack of conglomerates in these two areas suggests that the quarried strata are high in the formation. Similar strata crop out east of the Lickey Hills, and a borehole at Longbridge Pumping Station [0072 7755] records 89.15 m of sandstone, with five mudstone partings up to 0.9 m thick in the lowest 51.7 m. The total thickness proved here is almost certainly reduced by faulting, and the borehole is some distance from the Bromsgrove Sandstone outcrop. The full thickness of the Kidderminster Formation here may be as great as at Brockhill, providing no evidence for an active positive structural feature in the Lickey Hills area at this time.

East of the main outcrop, the top of the Kidderminster Formation was just penetrated by the Longbridge Borehole, and it is presumed to persist eastwards across the Knowle Basin to the margin of the Warwickshire Coalfield. A seismic line east of Sedgemere [230 755] (Allsop, 1981) suggests the presence of about 625 m of Sherwood Sandstone, with an intra-Triassic reflector 375 m above the base. The strata above the reflector would be Bromsgrove Sandstone, and those below are best interpreted as Kidderminster Formation with the possible addition of the Wildmoor Sandstone, as the base of this formation is gradational and might not be imaged on a seismic survey. The seismic evidence suggests that the Kidderminster Formation dies out close to the Meriden Fault, and south of here it is absent from boreholes at Shrewley [2222 6755] and Heath End (Figure 6). The reason for its disappearance is the unconformity at the base of the overstepping Bromsgrove Sandstone, and it is not known how far to the south-east it formerly extended.

Wildmoor Sandstone

The name Wildmoor Sandstone was introduced (Warrington et al., 1980) for beds formerly termed Upper Mottled Sandstone. This formation consists predominantly of sandstone and provides the well-known moulding sands quarried around Wildmoor. It crops out in the north-west, and attains a maximum recorded thickness of 134 m at Webheath, but thins rapidly north-east of its main outcrop, and is not known in the east of the area.

SEDIMENTOLOGY AND MODE OF DEPOSITION

The Wildmoor Sandstone is dominated by remarkably uniform, very weakly cemented, fine-grained, silty, micaceous sandstone. A typical sample from Messrs. Stanley

Evans moulding sand pit at Wildmoor [955 760] contained 12 per cent clay, 81 per cent silt and sand finer than 0.2 mm, and 7 per cent sand of 0.5 mm size. This sandstone is a distinctive deep 'foxey' red colour, probably as a result of its grain size and of the iron oxide which coats the sand grains. The beds are planar-laminated with some bedding planes covered with white mica, or show planar or trough cross-bedding with sets 0.3–1.5 m thick (Plate 5). The rare red-brown and grey-green mudstone beds are generally only a few centimetres thick but a few attain 30 cm and, exceptionally, 1.5 m. Mudstone clasts and clay laminae are, however, commonplace. Pebbly bands with quartz pebbles up to 2.5 cm across occur only rarely and are found low in the sequence, where they are associated with the coarser sandstones. Slump structures, ripple marks and mudcracks are known, and patchy calcareous cement occurs sporadically in the sandstones.

The formation apparently includes upward-fining rhythms, which commence with a medium- to coarse-grained or pebbly sandstone, and pass upwards through cross-bedded, fine-grained sandstones into planar-bedded, fine-grained sandstone and mudstone; the coarser-grained and pebbly sandstones are usually absent, particularly in the upper part of the sequence. The rhythms are difficult to distinguish, except in borehole cores and weathered pit faces, because the overall fine grain size obscures both the sedimentary structures and the grain size variations.

Hassan (1964) measured palaeocurrent directions in the area, and from both trough and planar cross-bedding recorded a unimodal direction of transport from the south-east. Fleet (1923) and Hassan (1964) studied the heavy mineral suite and recorded abundant tourmaline and zircon, with apatite and anatase and small quantities of feldspar, garnet, rutile, baryte, staurolite, ilmenite and magnetite.

The mode of origin of the Wildmoor Formation is uncertain. The fine grain size and the high degree of rounding and sorting of the grains may indicate that the sand was originally of aeolian origin, but the presence of micaceous bedding

Plate 5   Wildmoor Sandstone showing cross-bedding at Vigo Bridge [9860 7133]. Near the base are two mudstone beds which have been selectively weathered and are green with algal growth. (A13757)

planes and the cross-bedding sets, with parallel foresets persisting for long horizontal distances, imply aqueous deposition. Fluvial deposition is suggested by the apparent presence of upward-fining cycles, the common large-scale cross bedding and the scarcity of fossils. The low proportion of mudstone and flat-bedded sandstone (overbank deposits) argue against a high sinuosity river, while the fine grain of the sandstones, the scarcity of conglomerates and the lack of crevassing argue against a low sinuosity one. On balance, low sinuosity fluvial deposition accords best with the sedimentary structures, and the unusually fine grain size may be explained by the source material available for transport. The model suggested is thus a series of seasonally active braided streams transporting material from areas of sand dunes a relatively short distance away to the south-east, rather than the major river system which was active during deposition of the Kidderminster Formation.

Palaeontology

No fossils have been found in the Wildmoor Sandstone in the district. However, trace fossils are known from the formation in boreholes at Bellington [877 768], some 8 km to the west, in the adjoining Droitwich (geological sheet 182) district (Wills, 1970b; Wills and Sarjeant, 1970). These comprise invertebrate traces (*Planolites*) and vertebrate tracks (*Hamatopus* sp.; regarded as *Rhynchosauroides* sp. by Demathieu and Haubold (1972)). Farther west in that district, a fragmentary fish was recovered from the formation at Aggborough [836 757], Kidderminster (White, 1950). This may be representative of a group, the perleidids, that had marine affinity, and its presence, in proximity to occurrences of tracks of land vertebrates, presents problems of environmental interpretation. Possibly, as suggested by White (1950), its fragmentary condition reflects predation or scavenging by land vertebrates and the specimen may be somewhat remote from its habitat.

Stratigraphy

The Wildmoor Sandstone rests conformably upon the Kidderminster Formation, from which it is distinguished by its fine grain and 'foxey' red coloration. Exposures of the base were not found during the current survey. It is seen to be gradational in boreholes and is taken at the base of the lowest significant fine-grained 'foxey' red horizon. The formation cannot be further subdivided, and Wills (1976) included the whole within the upper part of his miocyclothem BSIV.

Complete sequences of Wildmoor Sandstone are known from water boreholes at Washingstocks and Webheath (Figure 4). The thickness in the former is taken to be 108.5 m, although both the top and bottom limits are poorly defined, and the logs made by T C Cantrill and L J Wills disagree about both. In Webheath Boreholes Nos.1 and 3 (logged by B C Worssam) the base of the Bromsgrove Sandstone was clear and thicknesses of 117.3 m and 134.0 m respectively were measured. Most of the strata are cross-bedded, with some desiccation cracks, slumps and ripple marks, and there are three beds of sparsely conglomeratic coarse sandstones with quartz pebbles up to 2.5 cm across.

Other boreholes, recording incomplete sequences of Wildmoor Sandstone, are shown in Figure 4.

The main outcrop of the Wildmoor Sandstone extends from Madeley Heath to Brockhill, but owing to the softness of the strata this is low lying and largely drift covered. The sandstone is worked for moulding sand at two pits at Wildmoor, just inside the adjoining Droitwich district [953 769], where at least 30 m of remarkably uniform fine-grained strata are visible. Indistinct upward-fining cycles and large-scale cross-bedding can be detected in weathered faces, together with a few mudstone bands a centimetre or so thick, and partings with large black and white mica flakes.

East of the main outcrop the formation is cut out rapidly by the overstepping Bromsgrove Sandstone; small areas were mapped at Barnt Green [006 730; 008 736], but none was found at outcrop in Longbridge or in the Longbridge Laundry Borehole (Figure 6). Boreholes at Lifford record 6.4 m of soft, deep red, cross-bedded Wildmoor Sandstone beneath the Bromsgove Sandstone, but none was present to the east in boreholes at Shrewley [2222 6755] and Heath End (Figure 6).

Bromsgrove Sandstone

This name was introduced (Warrington et al., 1980, p.38) for a succession of sandstones with subordinate siltstones and mudstones that, unlike the underlying Sherwood Sandstone Group formations and most of the overlying Mercia Mudstone Group, is comparatively fossiliferous. As originally defined, the top of the formation corresponded with the top of the 'Bromsgrove Stage' of Wills (1970a), but in this account it has been placed somewhat higher, on grounds of mappability, and corresponds with the top of the 'Waterstones' as defined by Hull (1869). The formation, therefore, includes the 'Basement Beds', 'Building Stones' and 'Waterstones' of Hull (1869), and equates with the former 'Lower Keuper Sandstone' of the Central Midlands. Hull's (1869) names for the divisions of the formation have remained in use though their application to lithostratigraphical successions has been variable and sometimes vague. Warrington (1968, 1970) applied the terms to distinct lithofacies units which are here formalised as the Burcot (lowest), Finstall and Sugarbrook members of the formation; they are described largely from a detailed log of the core of the original (undeepened) Sugarbrook No.3 Borehole (Figure 5; Warrington, 1968).

The formation crops out in the north-west of the district, where resistant sandstone beds give rise to a strong topography, and in a very small area in the extreme north-east. Borehole and geophysical evidence suggests that it is also present beneath the Mercia Mudstone Group and younger strata throughout the district.

Sedimentology and mode of deposition

The Bromsgrove Sandstone comprises numerous upward-fining sedimentary cycles (Figure 5). In the lower part of the formation these consist largely of coarse sandstones with a basal conglomerate or breccia bed that rests upon an erosion surface. Upwards through the formation the overall grain-

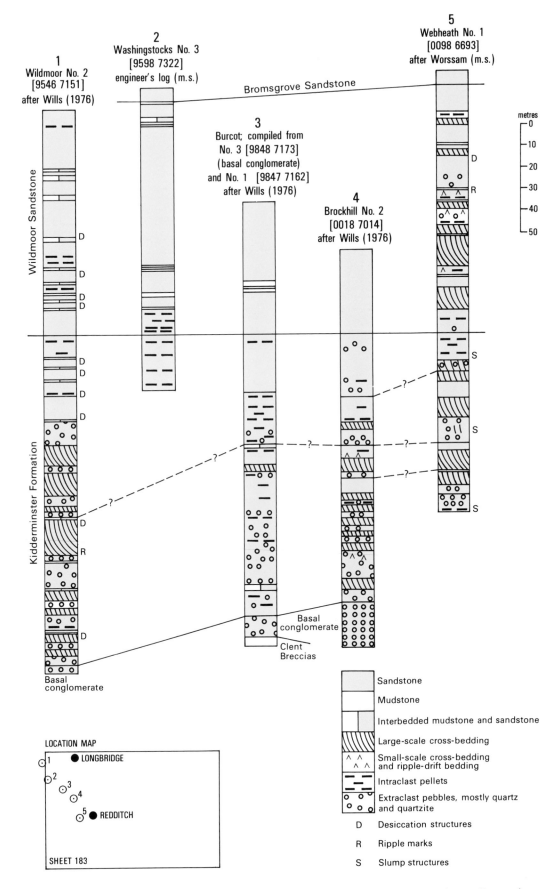

Figure 4 Borehole sections through the Wildmoor Sandstone and Kidderminster Formation

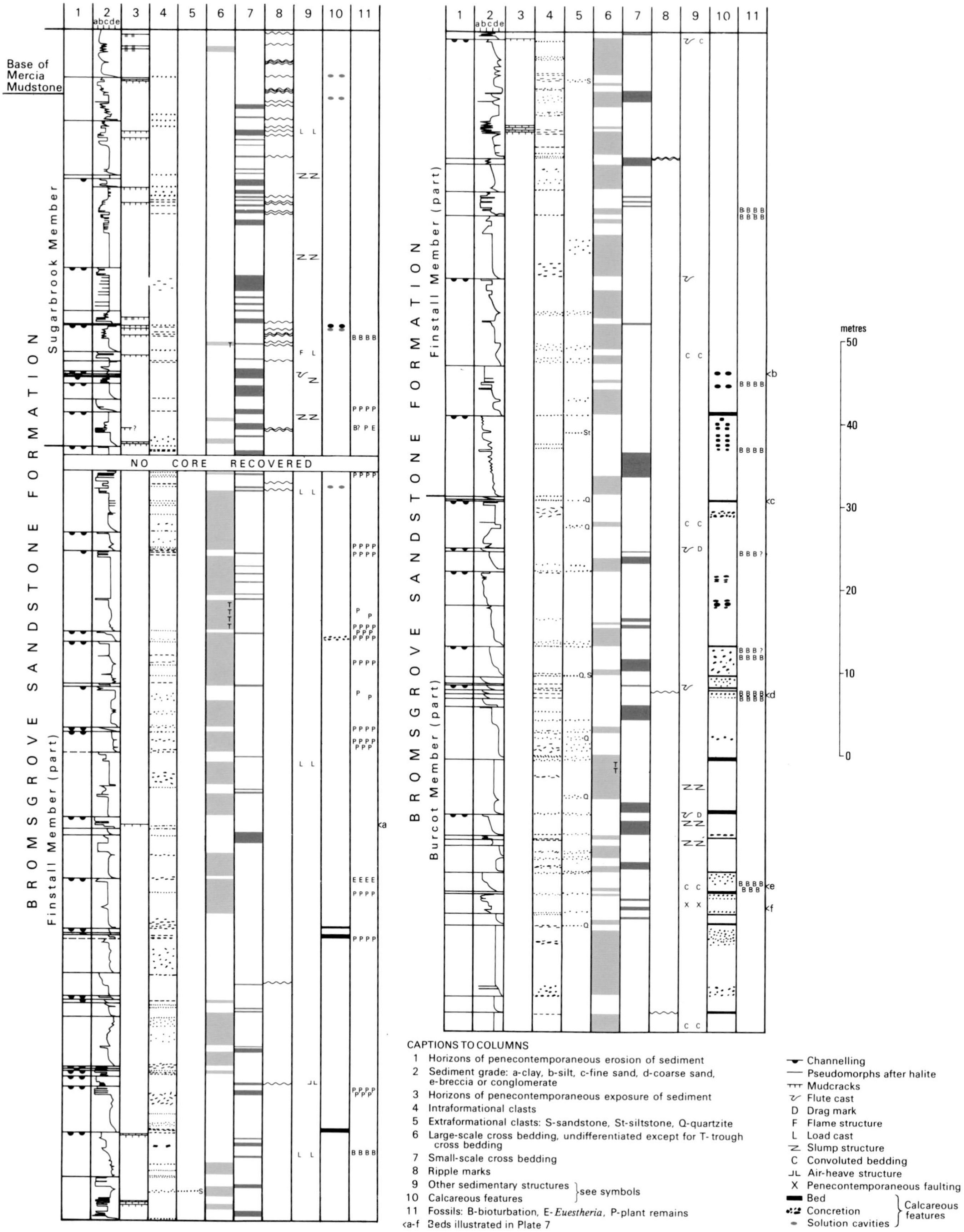

Base of Mercia Mudstone
BROMSGROVE SANDSTONE FORMATION
Sugarbrook Member
Finstall Member (part)
Burcot Member (part)
NO CORE RECOVERED
metres
50
40
30
20
10
0
CAPTIONS TO COLUMNS
1 Horizons of penecontemporaneous erosion of sediment
2 Sediment grade: a-clay, b-silt, c-fine sand, d-coarse sand, e-breccia or conglomerate
3 Horizons of penecontemporaneous exposure of sediment
4 Intraformational clasts
5 Extraformational clasts: S-sandstone, St-siltstone, Q-quartzite
6 Large-scale cross bedding, undifferentiated except for T-trough cross bedding
7 Small-scale cross bedding
8 Ripple marks
9 Other sedimentary structures
10 Calcareous features
}see symbols
11 Fossils: B-bioturbation, E-Euestheria, P-plant remains
<a-f Beds illustrated in Plate 7
Channelling
Pseudomorphs after halite
Mudcracks
Flute cast
D Drag mark
F Flame structure
L Load cast
Slump structure
C Convoluted bedding
Air-heave structure
X Penecontemporaneous faulting
Bed
Concretion
Solution cavities
Calcareous features

Figure 5 The sedimentary structures of the Bromsgrove Sandstone in the Sugarbrook No. 3 Borehole (after Warrington, 1968)

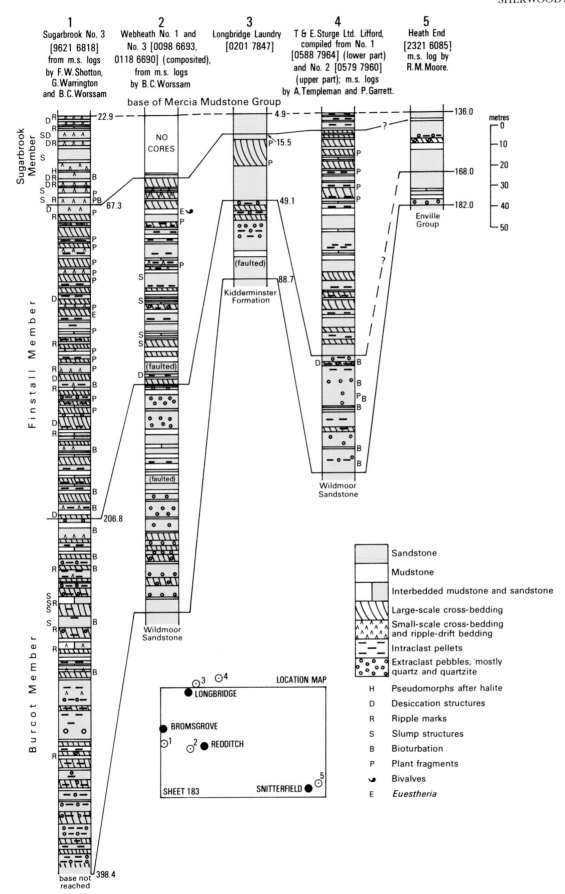

Figure 6 Borehole sections through the Bromsgrove Sandstone

size decreases, and siltstones and mudstones become more common components of the upper parts of the cycles. The presence of fresh feldspar pebbles and fragments, especially in the Burcot Member, distinguishes the Bromsgrove Sandstone from the older Triassic formations (Wills, 1948, p.53; 1970a, pp.249–250).

The *Burcot Member* (formerly the 'Basement Beds') is named from the Burcot area [98 71] where the base of the Bromsgrove Sandstone is mapped on Wildmoor Sandstone near the western margin of the district; this boundary is better seen to the west in the adjoining Droitwich (geological sheet 182) district (Mitchell et al., 1962). The internal character of the greater part of the member is illustrated in the logs of the Sugarbrook No. 3 Borehole (Figures 5 and 6) though, even when deepened, that borehole did not prove the base of the member. Upward-fining cycles in the member typically commence with a structureless polymict basal conglomerate bed that rests upon an erosion surface, and passes upwards into thick, cross-bedded and irregularly bedded or structureless red-brown sandstones. These lithologies dominate the Burcot Member cycles which include only minor beds of siltstone and mudstone (Plate 6).

The conglomerates contain subrounded to angular extra-formational clasts (up to 7 cm across) of quartz, quartzite, limestone, chert, quartz-tourmaline rock, tuff, sandstone, siltstone and feldspar, in addition to intraformational sandstone, siltstone and mudstone clasts. Mudstone clasts form tabular flakes and mud balls up to 10 cm across. The conglomerates commonly have a hard calcareous matrix and also contain calcareous and dolomitic concretions up to 10 cm across ('catbrain'). They pass up into coarse sandstones with angular grains, including fresh feldspar, which exhibit planar, lenticular and trough cross-bedding. These, in turn, grade upwards through finer sandstones, with small-scale cross-bedding (Plate 7f), sporadic ripple marks, and bedding planes covered with mica, into red-brown, chocolate, and grey-green mudstones. The latter are commonly absent from a cycle, but are represented by intraformational

Plate 6 Burcot Member of the Bromsgrove Sandstone, Pikes Pool Lane [9840 7097] showing massive and cross-bedded sandstone with thin mudstone beds and scattered pebbles. Sandstone from this quarry was used in construction of bridges on the Birmingham and Gloucester Railway. (A13759)

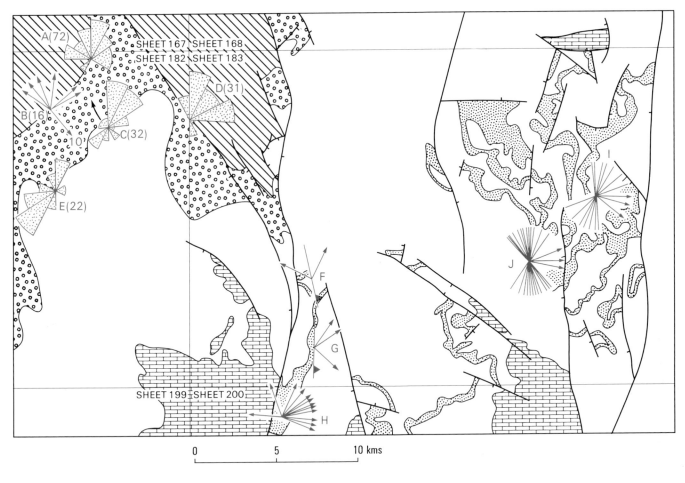

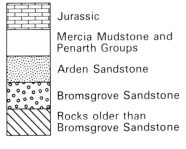

Jurassic

Mercia Mudstone and
Penarth Groups

Arden Sandstone

Bromsgrove Sandstone

Rocks older than
Bromsgrove Sandstone

A   Yielding Tree and Barnetthill [891 769]
B   Stone [863 748]
C   Chaddesley Corbett [892 749]
D   Catshill [955 734]
E   Elmley Lovett [869 696]
F   Hunt End [025 645]
G   Holberrow [028 614]
H   Quarrypits Farm, Inkberrow [006 563]
I   Rowington [201 690]
J   Henley-in-Arden [166 654]

Strike of ripple crests:

▲   lee side of ripple, where known,
       is shown by the triangle

Dip directions of current bedding foresets:

(72)   total number of readings at each locality

→   arrows show individual readings where
       there are fewer than 20 at a locality

where there are more than 20
readings, they are shown in 30°
sectors as a percentage of the total

A, B and D : Burcot Member
C and E   : Finstall Member
F to J   : Arden Sandstone Member

Figure 7   Current bedding foreset dip directions and ripple-crest strikes for the Bromsgrove
Sandstone and Arden Sandstone (after Warrington, 1968)

a

b

c

d

e

f

Plate 7 Sediments and structures in the Finstall and Burcot members of the Bromsgrove Sandstone Formation, Sugarbrook No. 3 Borehole, Bromsgrove. (Core levels illustrated are identified (a–f) on the graphic log of the borehole; Figure 5).

a   Finstall Member, c.45 m below top; brown micaceous shale and thin sandstones with mud-cracked clay laminae, mudflakes and mudstone intraclasts.

b   Finstall Member, 15 m above base; top of fining-upwards unit. Brown mudstones, silty mudstones and fine sandstones overlain by overbank mudstones with calcareous nodules ('catbrain'); scale in inches.

c   Burcot Member, 1 m below top; base of fining-upwards unit. Irregular eroded surface of brown mudstone overlain by intra-formational conglomerate (calcite-cemented) passing up into brown sandstones with sporadic quartzite pebbles.

d   Burcot Member, 24 m below top; base of fining-upwards unit. Light brown fine-grained sandstones, overlying an eroded surface on dark brown mudstone, are interbedded with siltstones and mudstones and show bioturbation and patchy calcareous cement; scale in inches.

e   Burcot Member, 47 m below top; upper part of fining-upwards unit. Brown fine-grained sandstone with mudstone laminae overlain by bioturbated sandy and silty mudstone with patchy calcareous cement.

f   Burcot Member, 50 m below top; disturbed small-scale cross-bedding and lamination in brown and patchily white fine-grained sandstone.

debris in the succeeding conglomerate. Flute casts, drag marks, slump structures and convolute bedding (caused by hydroplastic deformation) occur in the sandstones.

Measurements made on large-scale, cross-bedded sandstone units near Catshill, and near Barnetthill in the Droitwich (geological sheet 182) district (Figure 7; Warrington, 1968) indicate a north-easterly current flow during deposition of the member, and this is supported by a limited number of measurements made during the present survey.

A fine-grained sandstone from near Burcot [9832 7089], examined in thin section by Mr G E Strong, comprises moderately sorted, angular and subangular quartz grains with some white mica, biotite, clay pellets, opaque grains, intergranular and some detrital gypsum, authigenic clay minerals and secondary iron staining. Hassan (1964), following Fleet (1923), studied the accessory detrital minerals from this area and recorded fairly fresh feldspar (orthoclase, microcline and oligoclase), abundant tourmaline and zircon (the last occurring as less-rounded grains than in the Wildmoor Sandstone), and staurolite. Apatite, anatase, ilmenite and magnetite are less common than in the Wildmoor Sandstone but garnet is more plentiful.

The sedimentary structures and the nature of the cycles in the Burcot Member indicate deposition under fluvial conditions, with regular variation of current intensity and water depth. The basal conglomerates represent channel lag deposits; the overlying cross-bedded sandstone sets formed in large migrating sand bodies within the channel; and the succeeding fine sandstones and mudstones represent overbank deposits. However, the last are poorly represented in the member, suggesting deposition in a low-sinuosity or braided distributary system rather than a meandering one

(Warrington, 1970). This interpretation is supported by the coarseness of the deposits, the presence of channel lag deposits and intraformational clasts, the low variation in azimuth of the palaeocurrent directions (Figure 7) and the evidence of a high rate of sedimentation and a high degree of crevassing. As the braided stream deposits of the Bromsgrove Sandstone are noticeably more cyclic than those of the Kidderminster Formation and Wildmoor Sandstone, they are attributed to the Donjek type of Miall's (1977) classification, in which the cyclicity is explained either by sedimentation at different topographic levels within the channel system, or by phases of aggradation alternating with channel switching.

Calcareous nodules in the member (Plate 7d,e) probably represent caliche of pedogenic origin, formed during times of low water flow and abandonment of many stream courses. Despite the southerly origins indicated by the palaeocurrent measurements and the similarity of the pebble suite to that of the Kidderminster Formation, the lack of the pebbly facies in correlatives of the Bromsgrove Sandstone in southern England suggests (Audley-Charles, 1970) that pebbles of Armorican provenance in the member were not derived directly from that distant source, but were reworked from the Kidderminster Formation or its correlatives such as the Budleigh Salterton Pebble Beds, which are considered to have undergone erosion after the Hardegsen earth movements (Holloway et al., 1989). The presence of fresh feldspar, which is absent from the Wildmoor Sandstone, implies the emergence of a fresh source of plutonic rock. Thus, coarse and fine fractions in the member probably include material of both local and more distant southerly provenance.

The *Finstall Member* (formerly the 'Building Stones') is named from the Finstall area [980 700] to the east of Bromsgrove. Exposures are scarce in the district but are more extensive farther west in the Droitwich (geological sheet 182) district (Mitchell et al., 1962), where an important biota was recovered from the member in quarries at Rock Hill [948 698], some 3 km west of Finstall (Wills, 1910a). The internal character of the member is illustrated in the log of Sugarbrook No. 3 Borehole (Figures 5 and 6). The boundary with the Burcot Member is here placed at the base of the cycle that follows the highest occurrence of quartz or quartzite extraclasts.

The fining-upward cycles in the member broadly resemble those of the Burcot Member. Each rests upon an erosion surface that is commonly overlain by an intraformational lag conglomerate, which passes up into thick medium- to coarse-grained sandstone units (Figure 5). The sandstones are pale red-brown, buff or greenish grey, display large-scale planar, lenticular and trough cross-bedding, and were formerly widely used for building stone.

In the lower part of the member, 'catbrain' and sandstone and siltstone extraclasts (but not quartz or quartzite) are present (Figure 5). Fine-grained, flat-bedded sandstones and mudstones are more commonly developed towards the tops of cycles than in the Burcot Member. Bioturbation occurs in the lower part of the member, and plant remains and crustacea (*Euestheria*) in the higher part. Ripple marks, small-scale cross-bedding, flute casts, load casts, air-heave structures, convolute bedding and mudcracks (Plate 7a) are scarce, and slumps and drag marks are absent (Figure 5).

Measurements made on large-scale, cross-bedded sandstone units in the member at Chaddesley Corbett and Elmley Lovett, in the Droitwich (geological sheet 182) district, display greater variation of palaeocurrent directions than those from the Burcot Member (Figure 7; Warrington, 1968).

A sandstone from Finstall [9763 7030], examined in thin section by Mr G E Strong, comprises detrital quartz, some feldspar (microcline), rounded detrital gypsum grains and a few clay pellets, with some iron staining and secondary euhedral gypsum crystals.

The sedimentary structures and the nature of the rhythms, which are not greatly different from those in the Burcot Member, indicate the Finstall Member to be fluvial in origin. However, the greater variability of the palaeocurrent vectors and the finer overall nature of the sediments, with a greater proportion of fine-grained overbank sandstone and mudstone, suggest meandering rather than braided streams (Warrington, 1968, 1970; Selley, 1970). The presence of plants and of fresh to brackish water crustacea (*Euestheria*) indicates the existence of levées and tranquil overbank floodplain environments.

The *Sugarbrook Member* (comprising the former 'Waterstones') is named from the Sugarbrook pumping station site [961 681] to the south of Bromsgrove. Exposures of the member are scarce and its internal character is best known from the log of the Sugarbrook No.3 Borehole (Figures 5 and 6). The boundary with the Finstall Member is here placed at the top of the highest significant (>1 m thick) bed of sandstone with large-scale cross-bedding; core loss occurred around this level (at 67.3 m) in the borehole. The top of the member, at the boundary of the Bromsgrove Sandstone Formation (and the Sherwood Sandstone Group) with the overlying Mercia Mudstone Group, is here placed, on grounds of mappability, at the top of the highest significant grey sandstone forming the top of the 'Passage Beds' unit in the higher part of the former 'Waterstones'. At Sugarbrook this is some 9 m higher than the boundary suggested by Warrington et al. (1980, p.39), who included the 'Passage Beds' in the Mercia Mudstone.

Fining-upward cycles in the member are generally finer grained than those in the Finstall Member and are more complex, with a greater number of thinly bedded and interbedded sandstones, siltstones and mudstones; small-scale channelling and other evidence of erosion are present at the base of most cycles (Figure 5). The sandstones are medium to fine grained and display ripple-drift and small-scale cross-bedding, ripple-marked surfaces, slump structures, flute casts, flame structures, load casts and mudcracks; bedding planes are commonly mica covered. Large-scale cross-bedding is virtually absent but intraformational conglomerates occur throughout (Figure 5). Plant remains and crustacea (*Euestheria*) occur near the base of the member and bioturbation is also present in the lower beds. Pseudomorphs after halite occur above the level of bioturbation (Figure 6). The sedimentology of the Sugarbrook Member, and the presence in it of marine microplankton (acritarchs), indicate subaqueous deposition in an estuarine or littoral marine environment (Warrington, 1968, 1970).

PALAEONTOLOGY

The Bromsgrove Sandstone Formation has, in its type area, yielded one of the richest and most diverse floral and faunal associations known from British Triassic successions below the Penarth Group. The majority of these remains are from the Finstall Member; some have been recovered from the formation within the district, but the most extensive collections were made by L J Wills in quarries at Rock Hill [948 698], less than 1 km to the west of the district (Wills, 1907a, b, 1910a, b, 1916, 1947). Though noted in the memoir for that area (Mitchell et al., 1962), they are included here, with more recent additions and revisions to the biota, because of their significance for environmental interpretation.

Plant remains were noted on the western edge of the district by Murchison and Strickland (1840). The flora subsequently collected there from the Finstall Member by Wills comprises *Equisetites arenaceus?*, roots, pith casts and leaves of *Schizoneura paradoxa*, *Chiropteris digitata?*, leaves and stems of *Yuccites vogesiacus*, pith casts and decorticated stems of *Voltzia?*, an obscure cone *(Strobilites* sp.) and coniferous wood (Arber 1909; Wills, 1910a, b). A cone assigned to *V. heterophylla* was renamed *Masculostrobus willsi* by Townrow (1962); another specimen, identified by Grauvogel-Stamm (1969) as *M. rhomboidalis*, was reassigned (Grauvogel-Stamm, 1972) to *M. bromsgrovensis*. In further revision of this material (Grauvogel-Stamm and Schaarschmidt, 1978, 1979), *M. bromsgrovensis* and *M. willsi* were assigned to a new genus, *Willsiostrobus*, and a probable association with leaves of *Yuccites* was noted. An additional conifer, *Aethophyllum*, is represented in the Bromsgrove material in the British Museum (Natural History) (C R Hill, personal communication).

Schizoneura?, *Voltzia?* and *Yuccites* were recorded at Longbridge (Robertson and McCallum, 1930) and *Schizoneura* occurs in the Webheath boreholes. In boreholes in the district (Figure 6), plant remains, including spores and pollen, have been recovered almost exclusively from the Finstall Member.

The plant macrofossil association comprises equisetalean pteridophytes and coniferalean gymnosperms. Microfloras (Warrington, 1970) reflect the existence of a more diverse parent flora, including lycopsids, sphenopsids, pteropsids and gymnosperms; the last, principally conifers but including cycadalean types, dominated the flora. Wills' (1910a) illustration of spores and pollen from fructifications from Bromsgrove was only the second occasion that Triassic palynomorphs were recorded. Couper (1958), Townrow (1962) and Grauvogel-Stamm (1972) carried out further studies of in-situ pollen from *Willsiostrobus*, revealing the affinities of several forms of bisaccate gymnosperm pollen that dominate dispersed pollen associations from the formation (Clarke, 1965; Warrington, 1970; Pattison et al., 1973).

Acritarchs and tasmanitid algae were recovered from the Sugarbrook Member of Sugarbrook No.1 Borehole (Warrington, 1967, 1970).

The fauna of the formation comprises annelids, molluscs, arthropods and vertebrates. The annelid *Spirorbis* occurs attached to plant remains (*Schizoneura paradoxa*) (Ball, 1980). Small bivalves were recorded by Wills (1910a) as *Mytilus? sphinx*; Pattison et al. (1973) considered these might be referable to *Modiolus*, though Forbes (*in* Rose and Kent, 1955) suggested Wills' attribution was appropriate.

Arthropods comprise representatives of the Scorpionida and Crustacea. The scorpion genera *Mesophonus,* with six species, and *Spongiophonus* (monospecific), were introduced by Wills (1910a, 1947). Kjellesvig-Waering (1986) revised this material, retaining *Spongiophonus* and one species of *Mesophonus,* with a second questionably attributed to that genus, and introducing the new monospecific genera *Bromsgroviscorpio* and *Willsiscorpio.* Crustacea comprise the branchiopod conchostrachan *Euestheria minuta* which is known principally from the Finstall Member in the Sugarbrook (Figure 8a) and Webheath boreholes.

Vertebrates are represented by fish, amphibians and reptiles. A fish spine, assigned to the selachian genus *Acrodus,* is recorded from the Burcot Member near Tutnall (Wills, 1907a, 1910a). The Finstall Member has yielded the lungfish *Ceratodus* (Wills, 1910a), and *Dipteronotus cyphus,* a perleidid actinopterygian (Plate 8; Egerton, 1854; Wills, 1910a; Gall et al., 1974).

Amphibian remains from Bromsgrove (Wills, 1910a, 1916) have been reassigned by Paton (1974) to the capitosauroid labyrinthodont taxa *Cyclotosaurus pachygnathus* and *Mastodonsaurus lavisi.* Reptiles are represented by remains of the rhynchosaur *Rhynchosaurus brodiei* Benton (Benton, 1990), the rauisuchian *Bromsgroveia walkeri* Galton (Galton, 1985), a lepidosaur related to *Macrocnemus,* and a nothosaur (Walker, 1969).

Burrows and bioturbation noted in the Bromsgrove Sandstone in Sugarbrook No. 3 Borehole (Figures 5, 6; Plate 7) occur principally in the upper part of the Burcot Member and the lower parts of the Finstall and Sugarbrook members.

ENVIRONMENTAL INTERPRETATION

Pteridophyte plants probably grew in damp tracts bordering river channels or in floodplain areas, mainly during deposition of the Finstall Member. Wills (1910a, b) noted stems of *Schizoneura paradoxa,* apparently in growth position, and rootlets, and suggested that this plant may have grown in water. The gymnosperm components of the flora probably occupied drier habitats. Scorpions indicate the presence of dry terrestrial habitats, which were probably 'enlivened by their chirping or hissing' (Wills, 1947); but the crustacea reflect the existence of seasonal pools of warm, fresh to brackish water. *Euestheria* from near the top of the Finstall Member in Sugarbrook No.2 Borehole are large (Figure 8a) and well preserved; many have complete chitinous carapaces which mostly taper posteriorly, though some are rounded in outline (Figure 8a), features comparable with sexual dimorphism shown by modern forms (Kobayashi, 1954) in which the rounded forms are females. Growth lines are closely spaced, and in some specimens become very closely spaced towards the carapace margin; in modern forms these features reflect optimal ecological conditions and maturity respectively (Kobayashi, 1954). This assemblage, including complete specimens of various sizes and probably both sexes (Figure 8a), some of which may have reached maturity, is interpreted as a biocoenosis that reflects directly the aqueous environments on the Finstall Member floodplain at Bromsgrove.

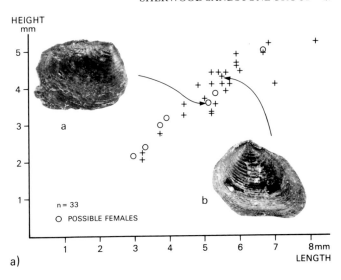

a)

Figure 8a Size variation in carapaces of *Euestheria minuta* from the Bromsgrove Sandstone in Sugarbrook No. 2 borehole (depth 68.88 m), Bromsgrove, Worcestershire. Specimens illustrated: Birmingham University Museum collection, a—BU 2132 (possible female), b—BU 2133 (possible male); ×5.

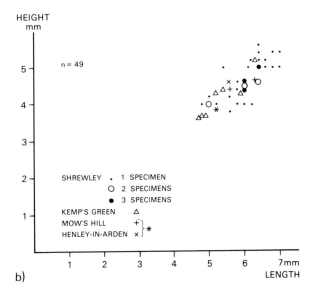

b)

Figure 8b Size variation in carapaces of *Euestheria minuta* from the Arden Sandstone. (All Shrewley examples on British Geological specimen GSM 118455; all others from L J Wills collection, Birmingham University Museum).

The presence of lungfish suggests that rivers on the Finstall Member floodplain were prone to seasonal drought but were intermittently penetrated, from the north (Warrington and Ivimey-Cook, in press), by marine-sourced water which introduced those and other fish, the bivalves, and semi-aquatic reptiles (lepidosaurs and nothosaurs) with marine affinity. Capitosauroid labyrinthodonts indicate the existence of at least seasonal freshwater bodies on a floodplain that was also inhabited by herbivorous reptiles (*Rhynchosaurus*) and carnivores.

Plate 8 *Dipteronotus cyphus* from the Finstall
Member, Bromsgrove Sandstone
Formation, Bromsgrove,
Worcestershire, ×1½. (Holotype,
British Geological Survey:
a—GSM 18188 b—GSM
18189; counterparts).

The influence of a marine environment in the succeeding Sugarbrook Member is reflected in a decline in the remains of terrestrial biota, and by the appearance of acritarchs.

STRATIGRAPHY

The Bromsgrove Sandstone rests variously upon Wildmoor Sandstone, Kidderminster Formation and Enville Group rocks. There is probably a stratigraphical break beneath the formation throughout the area, although there is no obvious angular discordance where it rests upon Wildmoor Sandstone in the Bromsgrove district. This break may be equated with the Scythian Hardegsen Disconformity in Germany (Trusheim, 1963; Warrington, 1970). After deposition of the Wildmoor Sandstone, uplift and erosion of the source areas in the Armorican mountains of north-west France and the English Channel resulted in deposition of the Burcot Member in a braided river system flowing northward through the Worcester and Knowle basins. The Burcot, Finstall and Sugarbrook members may be interpreted as the proximal, medial and distal portions of a subaerial delta complex, with the three lithofacies migrating southward as the source area was denuded and the transporting power of river system waned. These lithofacies are inevitably diachronous, but this is not apparent within the confines of the Redditch district where they span the interval late Scythian to early Ladinian (Warrington, 1970).

The Bromsgrove Sandstone and its component members vary considerably in thickness (Figure 6), apparently as a result of contemporaneous growth faulting. The region of thickest accumulation, around Bromsgrove, lies within the Worcester Basin, which was subsiding relatively rapidly (Wills, 1970a). Geophysical evidence suggests that relatively thick deposits are also present to the north-east within the Knowle Basin, but boreholes in the south-east and around Longbridge (Figure 6) show thin sequences on the margins of the positive areas of the Warwickshire Coalfield and the Lickey Axis respectively. The latter cannot, however, have been an absolute barrier, as the north-eastward palaeocurrent direction in the Burcot Member at Catshill (Figure 7) implies free flow across it.

The members within the Bromsgrove Sandstone have not been mapped in detail because their identification relies heavily on sedimentary structures which can only be seen in the better exposures; thus the outcrops shown in Figure 3 are generalised. The subdivisions are generally easily recognisable in boreholes (Figure 6) although some of the old records are inadequate for this purpose.

West of the Lickey End Fault the Burcot Member, similar in lithology to that proven at Sugarbrook, has a large outcrop north of Bromsgrove. Boreholes at Washingstocks and Barnsley Hall Hospital proved 121.8 m and 20.1 m respectively of this unit resting upon Wildmoor Sandstone. Pebbles of quartz and quartzite up to 7 cm across are recorded in sandstones at outcrop. Of two thick mudstone beds mapped just within the adjoining Droitwich district, that south-west of Barnsley Hall Hospital [954 720] may correlate with one low in the member in the Washingstocks boreholes, and that west of the hospital [955 725] may be one of two thick mudstone beds recorded high in the member at Sugarbrook (Figure 6).

East of the Lickey End Fault, the base of the Burcot Member is visible where the railway crosses the road between Tutnall and Burcot [9861 7132]; sandstone in a nearby exposure [9835 7137] contains fresh feldspar clasts, quartz pebbles up to 7 cm in diameter and rounded mudstone clasts up to 10 cm across. To the south, a fault cuts out a large part of the member, and the pebble-free sandstones cropping out at Tardebigge are assigned to the Finstall Member. The Sugarbrook Member, which crops out between Tardebigge and Webheath, comprises micaceous, fine-grained sandstones and mudstones, with less massive sandstone than that west of the Lickey End Fault, and is distinguishable from both the Finstall Member and the Mercia Mudstone by augering. Two water boreholes at Webheath (Figure 6) prove at least 244 m of Bromsgrove Sandstone, but the succession there is disturbed by faulting. The Sugarbrook Member was not cored. The underlying Finstall Member, which is faulted against Burcot Member at 73.8 m in the No. 1 Borehole, yielded Schizoneura, spores and pollen, Mytilus sphinx and Euestheria. In the No. 3 Borehole the Finstall Member apparently extends from surface to a depth of 97.5 m, and overlies 114.5 m of the Burcot Member. The top of the Bromsgrove Sandstone in the Putcheons Farm Borehole [0694 6577] is taken here at 148.1 m, 3 m higher than the position given by Poole (1969). The 18 m of Bromsgrove Sandstone in the borehole consists of alternating thin beds of cross-bedded brown sandstone and red-brown mudstone of Sugarbrook Member type with ripple marks, mud cracks and slump structures.

The Bromsgrove Sandstone may be at its thinnest along the axis of the Lickey Hills, although the true thickness in the narrow outcrop south of Barnt Green is unknown because of faulting. The Burcot Member is exposed in an old quarry adjacent to the railway [0110 7344], and boreholes near Cock's Croft [007 729] proved interbedded mudstones and fine- to coarse-grained, very micaceous sandstones, which are probably in the Sugarbrook Member.

Between the Barnt Green and Longbridge faults, the Bromsgrove Sandstone is well known from widening of the railway cutting (McCallum, 1927). Robertson and McCallum (1930, p.44) estimated the total thickness there to be 65.5 m, and their section may be interpreted as follows: Sugarbrook Member 9.1 m; Finstall Member, 25.9 m; Burcot Member, 30.5 m; the last forms a much smaller proportion of the Bromsgrove Sandstone here than is normal, and this may reflect a lower rate of subsidence along the Lickey Ridge relative to the surrounding areas. Most of the cutting is in the Burcot Member, which is repeated by complex faulting and gentle folding, but the Finstall Member is exposed at Cofton Richards Farm [0163 7564] and the Tower House [0168 7529], and yielded fossil plants during the railway widening [0122 7503]. East of the Longbridge Fault the Sugarbrook and Finstall members crop out across an anticline which may also include the Burcot Member at its core; the intraformational boundaries are uncertain here. This is also true in the outcrop of Bromsgrove Sandstone north of Longbridge, which is badly faulted and largely built-over. However, coarse Burcot Member sandstones with pebbles of quartz and yellow chert are exposed in Hanging Lane [0121 7907]; the Finstall Member, comprising false-bedded, brown sandstones with plant remains, is seen at the Pigeon House Hill road cutting [0174 7893]; and the Sugarbrook Member is revealed in the River Rea section north of Longbridge Lane [014 777]. Longbridge Laundry Borehole (Figure 6) proved a complete 83.8 m-thick sequence of the Bromsgrove Sandstone; although not detailed, the log indicates the following thicknesses: Mercia Mudstone 4.9 m; Sugarbrook Member 10.6 m; Finstall Member 33.6 m; Burcot Member 39.6 m.

The Sturge boreholes, Lifford, provide the most easterly location at which the three subdivisions of the Bromsgrove Sandstone can be clearly distinguished (Figure 6). The correlation between the boreholes adopted here follows Old (1983) and differs from that of Wills (1976). Borehole No.2 proved 8.2 m of Sugarbrook Member at 185.0 to 193.2 m depth, and Borehole No.1 proved 58.3 m of Burcot Member at 303.5 to 361.8 m depth, giving a thickness of about 111 m of Finstall Member, and a total of about 178 m of Bromsgrove Sandstone.

On the eastern edge of the area boreholes at Heath End and Shrewley (Wills, 1976) proved 46.5 m and 46.2 m of Bromsgrove Sandstone respectively, underlying Mercia Mudstone and resting on Enville Group, while boreholes at Claverdon [1985 6477] (Richardson, 1928) and Rowington [2090 6882] proved 14.3 m and 69.5 m respectively without penetrating the base. The best recorded sequences, at Heath End and Shrewley, are dominated by buff to pale grey-green massive, micaceous sandstones, with large-scale tabular cross-bedding and mudstone pellet conglomerates, suggestive of the Finstall Member. The quartz pebbles characteristic of the Burcot Member are recorded low in the sequence and the formation passes up into the Mercia Mudstone via interbedded mudstones and thin sandstones with ripple marks. It is suggested, therefore, that the three members recognised in the Bromsgrove area persist to the east. In the north-east of the area the Bromsgrove Sandstone crops out on the east side of the Meriden Fault, and seismic evidence (Allsop, 1981) indicates that it thickens westward across that fault, from about 60 m to about 250 m.

Wills (1970a, 1976) divided the 'Basement Beds', 'Building Stones' and 'Waterstones' into eleven 'miocyclothems' designated KSI-IV, KSV-IX and KSX-XI respectively, which have not been adopted in this account. On lithofacies definitions adopted here, part of miocyclothem KSV falls in the Burcot Member, and part of KMI (the lowest in Wills' 'Keuper Marl') is included in the Sugarbrook Member.

MERCIA MUDSTONE GROUP

This term was formally introduced (Warrington et al., 1980) for beds previously termed 'Keuper Marl'. The group consists predominantly of red-brown, blocky, unfossiliferous mudstones, that are generally poorly exposed. Thin, impersistent 'skerries' of sandstone or siltstone occur throughout the group; two of them, the Weatheroak and Arden sandstones, form thick, mappable units. The Droitwich Halite Formation probably occurs at depth in the south-west of the district, whereas gypsum characterises the upper part of the group at outcrop. Pale grey-green mudstones and siltstones of the Blue Anchor Formation cap the Mercia Mudstone succession.

The general succession has been compiled from a small number of well-documented boreholes and outcrop sections, and by detailed mapping of the few distinctive lithostratigraphical units. The group rests with apparent conformity upon the Bromsgrove Sandstone, and it is overlain by the Penarth Group; it almost certainly continues beneath the younger rocks throughout the district. Although complete sections are few (Figure 9) it is clear that the group thickens westwards from about 240 m in the east to more than 450 m at Saleway, 2.5 km to the west of the district.

Fossils are rare except in the Arden Sandstone Member, pollen from which are of late Carnian age (Warrington et al., 1980), constraining the Mercia Mudstone below to a Ladinian to mid-Carnian age, and that above to a Norian or younger Triassic age. No palynomorphs have been recovered from the group below the Arden Sandstone in the district, but sporadic specimens from the higher beds in the Knowle Borehole (Figure 13) indicate a late Triassic, Norian to Rhaetian, age for the Blue Anchor Formation and immediately underlying beds.

STRATIGRAPHY

The basal few metres of the Mercia Mudstone Group show an upwards passage from the Bromsgrove Sandstone (Figure 5) and consist of thin-bedded sandstones, siltstones and mudstones (Poole, 1969; Wills, 1976). The succeeding beds, below the Weatheroak Sandstone (see below), are predominantly argillaceous and some display cyclicity, as formerly seen in brickworks at Redditch [033 672] and Studley [063 637] (Jackson, 1982). Here, each cycle comprises thin, light grey to brown, dolomitic sandstones, locally with irregular bases, that display slump structures, small-scale cross bedding, ripple marks, and interlaminated siltstones and mudstones, succeeded by thicker, largely structureless, blocky, light to medium red-brown, silty, dolomitic mudstones with scattered quartz grains and green spots; these pass upwards into harder, brittle, more compact, dark red-brown mudstones with fewer green spots, less dolomite and more fine quartz grains, at the top of which a zone of secondary green coloration occurs below the coarser bed at the base of the succeeding cycle. Laminated units in such sequences are commonly colour-banded (red-brown and green), and mudcracks, mudstone intraclasts and pseudomorphs after halite occur at some levels in the sequence.

The Droitwich Halite Formation proved in the Droitwich area (Mitchell et al., 1962), and at Saleway (Wills 1970a, 1976; Poole and Williams, 1981), may extend at depth eastwards into this district as far as the Stoke Pound and Lickey End faults or beyond. Its correlation with the non-saliferous beds of the lower part of the group elsewhere is uncertain, but it may equate with the Spernall Gypsum (see below). A 20 m well at Morton Hall Farm [c.018 593] (Richardson 1930, pp.164, 194) and another to 69 m at Morton Hall [0186 5939] encountered saline water in mudstone below the Arden Sandstone.

Apart from the Weatheroak and Arden Sandstone members and the Blue Anchor Formation, no persistent lithostratigraphic marker units have been recognised in the group. The sequence below the Arden Sandstone is insufficiently documented to allow correlation with most of the formations recognised in the Mercia Mudstone elsewhere in

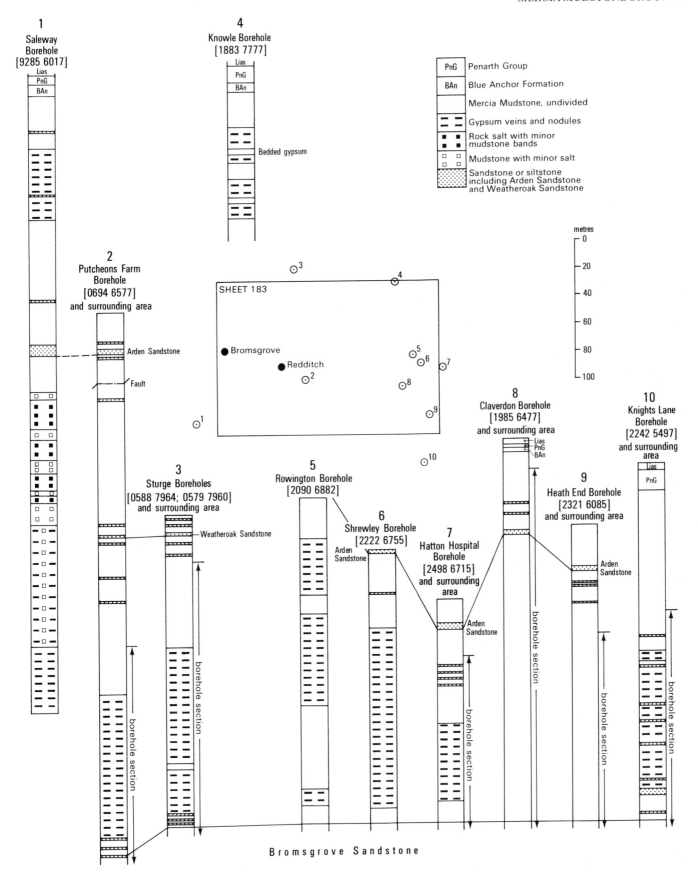

Figure 9 Comparative sections in the Mercia Mudstone

the English Midlands (Elliott, 1961; Warrington et al., 1980; Old et al., 1987). By comparison with Elliott's (1961) sequence the Weatheroak Sandstone may equate with the Cotgrave Skerry at the base of the Edwalton Formation, and the Arden Sandstone may mark the top of that formation. In the Knowle Borehole the Blue Anchor Formation rests directly on gypsiferous red mudstones correlatable with those of the Trent Formation, and there is locally no development of the Glen Parva Formation which elsewhere intervenes between the two.

Abundant thin gypsum veins and gypsum or anhydrite nodules are recorded in the lower part of the Mercia Mudstone Group in boreholes at Saleway (Poole and Williams, 1981) and Knowle (IGS, 1982, p.3); these are also present towards the top (Figure 9). A 3.5 m bed of coalescing gypsum/anhydrite nodules and secondary fibrous gypsum veins occurs at Knowle. Gypsum has been proved much less commonly in the middle of the group although it is abundant locally, for example at Spernall where it was formerly mined (p.58). The workings were in a bed lying about 25 m below the Arden Sandstone, and the BGS Spernall Borehole [1096 6231] proved 8.5 m of mudstone containing abundant veins of secondary fibrous gypsum, up to 8 cm thick, and a few tiny gypsum nodules. These beds passed down into mudstone with very little gypsum at 28 m below the Arden Sandstone.

Green reduction spots are commonplace in the cycles noted above, but it is not known whether these have a systematic distribution within the group as a whole. In the Knowle Borehole, a few black radioactive nodules, up to 1 cm across with reduced mudstone haloes up to 5 cm in diameter, were observed (Harrison et al., 1983). Some nodules contain coffinite cores with outer margins of chalcocite and niccolite: others are enriched in vanadium oxides and hydroxides.

The mineralogy of the Mercia Mudstone Group has been studied from a number of sites in the district (Jeans, 1978). Dolomite or calcite form up to 30 per cent of individual beds in both mudstones and 'skerries', and are considered by Jeans to have an evaporitic origin. Clay minerals identified include chlorite, corrensite, sepiolite and smectite.

The thin fine-grained sandstone and siltstone beds (skerries) that form the basal beds of cycles in the group are mostly of limited extent, and cannot be used to subdivide the succession. Two distinctive sandstone members are, however, more widely developed. The Weatheroak Sandstone has been traced from Longbridge [02 76] to Redditch [07 66]. It is a pale grey, flaggy, typically cavernous sandstone, with green mudstone interbeds normally less than 0.7 m thick. Copper mineralisation is present in the basal few centimetres at a number of localities. Secondary malachite predominates, mostly as grain coatings, but also as botryoidal growths in cavities. A particularly heavily mineralised specimen, west of Moorfield Farm [0639 7291], also contained cuprite with native copper inclusions and tenorite overgrowths, chalcocite with some marginal replacement to covellite, and a hydrous copper silicate, probably chrysocolla. Around Weatheroak Hill, the Moorfield Coppice Mudstone, a smooth, silty, distinctively laminated mudstone up to 0.3 m thick, occurs about 2 m below the Weatheroak Sandstone [059 743]. These beds proved devoid

of palynomorphs, and no other fossils have been observed.

The importance of the Arden Sandstone Member (Warrington et al., 1980) as a widespread stratigraphical marker within the Mercia Mudstone of the district, was demonstrated by Matley (1912). The Arden Sandstone, which locally gives rise to a prominent escarpment (Frontispiece), consists of grey and green sandstones, siltstones and mudstones, commonly finely interbedded and laminated, and with much bioturbation and numerous small-scale structures indicative of thixotropic movement of unlithified sediment. The sandstones and siltstones exhibit trough, planar and small-scale ripple drift cross-bedding. Massive sandstone beds occur in the thicker developments of the member (Plate 9) and have been much used for building purposes (p.58). Measurements made on cross-bedding and ripple marks indicate a generally easterly current flow during deposition of the member in the district (Figure 7). The member reaches 11 m in thickness but 3 to 8 m is more usual: comparative sections are shown in Figure 10, and the section at Shrewley canal cutting [21 67] is illustrated in Plate 10.

The Blue Anchor Formation consists of pale grey-green, blocky mudstone and siltstone. It varies in thickness from a maximum of 12 m in the south-east to only 10 cm at Round Hill (Figure 11), probably due to an unconformity at the base of the Penarth Group. Evidence for this unconformity is found in a few outcrops in the south-west where the top few centimetres of the formation are ferruginous. In the Knowle Borehole the base of the Westbury Formation interpenetrates the Blue Anchor Formation for 1 cm and contains fragments of the latter. The formation has yielded sporadic miospores and dinoflagellate cysts indicative of a late Triassic (Norian to Rhaetian) age (Figure 13), and fish scales.

PALAEONTOLOGY

The Arden Sandstone Member has yielded a relatively diverse flora and fauna (Plate 11). Plant remains comprising *Carpolithus* (Plate 11b), an equisetalean pteridophyte (*Schizoneura*) and coniferalean gymnosperms (*Voltzia* and ?*Yuccites*) were recorded by Wills and Campbell-Smith (1913), who also reported female cones of *Voltzia* (Plate 11a). Palynological residues contain a green alga (*Plaesiodictyon mosellanum*) and spores and pollen from a parent flora that included bryophytes, sphenopsids and pteropsids, but which was dominated by gymnosperms, mainly conifers but with some cycadalean types.

The fauna is represented by bivalves, crustacea, vertebrates and trace fossils. The bivalves (Plate 11c–f) include specimens that Newton (1893, 1894) tentatively placed in the genera *Nucula*, *Pholadomya* and *Thracia*. Cox (in Rose and Kent, 1955) was unable to regard them as definitely marine forms. The ribbing style and the shape of the '*Pholadomya*' does, however, compare with the marine taxon *Palaeocardita* cf. *austriaca*.

Crustacea comprise the branchiopod conchostrachan *Euestheria minuta*, typically represented by large specimens (Figure 8b). Vertebrate remains are largely those of fish. Teeth and spines of heterodontiform sharks (*Acrodus* [*Hybodus*] *keuperinus* (Plate 11g–i), *Phaebodus brodiei*) are common, but specimens of *Dictyopyge superstes*, a palaeoniscid

Plate 9 Arden Sandstone in canal cutting at Rowington [202 691]; the slumped unit at top is underlain by cross-bedded and planar-bedded units. (A13752)

(Plate 11j; Egerton, 1858) and *Semionotus brodiei* (Newton, 1887) also occur; McCune (1986) considers that *S. brodiei* shows some palaeoniscid characters and may not be a semionotid.

Land vertebrates are represented by cranial bones of labyrinthodont amphibians (Brodie, 1856, 1893), and by tracks of chirotherioid (Plate 11k) and rhynchosauroid type (Murchison and Strickland, 1840; Brodie, 1860; Sarjeant, 1974). Bioturbation is widespread and small burrows of *Planolites* type, and other trace fossils of invertebrate origin, are common (Plate 12).

The biota of the Arden Sandstone reflects terrestrial and subaqueous environments similar to those indicated by the flora and fauna from the Bromsgrove Sandstone Formation. The spores (*Porcellispora*) of bryophyte origin reflect damp terrestrial habitats that may also have been colonised by equisetalean plants and ferns. Water-laid sediments are extensively bioturbated. The green alga *Plaesiodictyon* is regarded as associated with brackish water environments (Wille, 1970) and numerous large *Euestheria* are indicative of optimal ecological conditions. Marine connection is indicated by sporadic bivalves and the widespread occurrence of shark remains.

Sparse pollen assemblages were recovered from the member at Hunt End [027 645] and Shurnock [028 603], in the south-west of the district. Richer palynomorph associations from farther east, at Matchborough Hill, Redditch [077 662], Lapworth Park [164 690], Copt Green [175 699] and Rowington [201 691], are mostly dominated by bisaccate gymnosperm pollen, including *Ovalipollis pseudoalatus* and *Ellipsovelatisporites plicatus*, but also contain *Porcellispora longdonensis*, *Duplicisporites granulatus*, *Camerosporites secatus*, *Vallasporites ignacii* and *Brodispora striata*, and are of late Triassic, late Carnian (Tuvalian) age.

Figure 10 Comparative sections in the Arden Sandstone

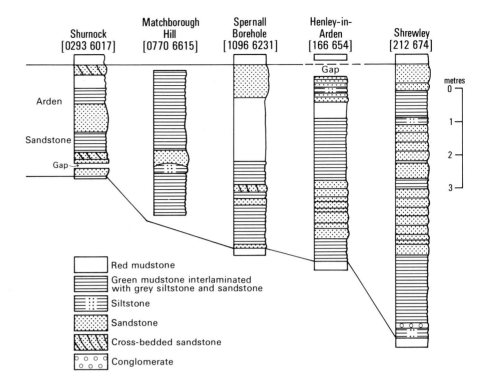

DEPOSITIONAL ENVIRONMENT

The predominantly red-brown colour of the Mercia Mudstone, coupled with the presence of mudcracks and evaporites, and a scarcity of fossils and laminated beds, has long been considered indicative of deposition in an arid environment. However, the upward passage from the estuarine or littoral marine Sugarbrook Member suggests that subaqueous environments, which were at least partly marine-sourced, persisted during deposition of the lowest part of the succeeding red mudstone succession. Within the Mercia Mudstone Group, the Arden Sandstone and, locally, the highest beds, including the Blue Anchor Formation, contain fossils indicative of marine environments, but such indications are lacking from the intervening evaporite-bearing mudstone sequences in the district. The cycles noted in those mudstone sequences are broadly analogous to ones documented by Arthurton (1980) in the group in Cheshire, in which the laminated lower and blocky upper units were interpreted, respectively, as having formed subaqueously and subaerially. Arthurton, following Glennie and Evans (1976), suggested that the sedimentology of the Cheshire sequences invited comparison with the present-day Ranns of Kutch, an arid coastal flat on the India–Pakistan border, that is subject to annual marine innundation and to local flooding by water of continental origin. Such analogy can only be an approximation, but the Mercia Mudstone sequences may be viewed as comprising fine terrigenous sediment transported by wind or water, and deposited partly subaerially and partly, in association with evaporites and neoformed clay minerals (Jeans, 1978), in shallow, low-energy, subaqueous environments. Connection to a marine environment is indicated by evidence from geochemistry, clay mineralogy and fossils; salinity would, however, vary with geographical situation and the relative rates of fluvial discharge, marine influx and evaporation (Warrington and Ivimey-Cook, in press).

Halite accumulations, such as the Droitwich Halite, represent periods when shallow bodies of steadily replenished, marine-sourced hypersaline water were subject to evaporation in basins that were subsiding rapidly in response to syndepositional faulting (Warrington, 1974). Concentrations of nodular sulphates in laterally equivalent mudstones represent precipitation from intrastratal brines in sabkha environments. Fibrous gypsum veins formed after lithification, but may reflect deposition from connate brines (Burgess and Holliday, 1974, pp. 22-23) rather than the alteration of primary nodular anhydrite to gypsum (Shearman et al., 1972).

The majority of the thin skerries in the mudstone sequences represent low-energy subaqueous conditions that were of short duration, and possibly seasonally or climatically related. The thicker Arden Sandstone Member, in contrast, represents a major environmental change, and an interruption of the conditions under which the greater part of the Mercia Mudstone accumulated. It is comparable with the Schilfsandstein (Wurster, 1964) of Germany. Dominantly arenaceous and thinner, more argillaceous sequences in the member are interpreted, respectively, as deposits of distributary channels and interdistributary areas in a deltaic or estuarine environment that developed in response to regional geographical changes (Warrington and Ivimey-Cook, in press).

The highest beds in the Mercia Mudstone Group, the Blue Anchor Formation, reflect changes that preceded the Rhaetian marine transgression in the district. Sporadic occurrences of the dinoflagellate cyst *Rhaetogonyaulax rhaetica* in these beds may record transient marine environments that were precursors of the main transgression which introduced the conditions under which the Westbury Formation of the

Plate 10 Arden Sandstone showing interbedded sandstones, siltstones and mudstones seen resting on red mudstone in the bottom right. Canal cutting at Shrewley, 150 m north-west of tunnel portal [2118 6744]. (A13530)

Penarth Group, with richer populations of *R. rhaetica* (Figure 13), formed.

PENARTH GROUP

This name was formally introduced by Warrington et al. (1980) for beds formerly termed 'Rhaetic'. The Penarth Group comprises deposits representing a marine transgression of late Triassic (Rhaetian) age, and its predominantly grey fossiliferous lithologies distinguish it from the earlier Triassic rocks. The constituent formations show variations of thickness and lithology (Figure 11) which reflect local contemporaneous earth movements, for example on the Vale of Morton Anticline at Round Hill (p.57). The group is subdivided (Warrington et al., 1980) into an older Westbury Formation and a younger Lilstock Formation. In this area the latter is represented almost entirely by the Cotham Member; the overlying Langport Member may be represented by a thin limestone at Knowle (Figure 11), but elsewhere in the district, if present at all, it comprises

mudstones which are indistinguishable from those of the overlying Blue Lias Formation.

Westbury Formation

The Westbury Formation consists largely of dark grey fissile mudstone with pale grey silty intercalations. Thin limestones occur sporadically and, in the south-west particularly, thin pale grey sandstones showing ripple drift and cross lamination occur near the base. The dark mudstones are locally pyritic, and pyrite encrusted fossil casts were encountered in the Knowle Borehole. The black pyritous mudstones of the Westbury Formation formed in shallow reducing marine environments, where current activity was sufficient to distribute only minor arenaceous influxes as lenticles and beds.

Between the Pennyford and Shelfield faults the Westbury Formation is locally very thin, and a trial pit at Round Hill (Figure 11) proved Cotham Member mudstones directly overlying a much attenuated Blue Anchor Formation.

Plate 11 Arden Sandstone fossils

a *Voltzia*; female fructification, ×2 (BU 2130, L J Wills collection; Shelfield?, Warwickshire)

b *Carpolithus*; ×1 (BU 2131, L J Wills collection; Shelfield?, Warwickshire)

c *Thracia? brodiei*; ×2 (GSM 4978; Shrewley, Warwickshire)

d *Nucula? keuperina*; ×2 (Holotype: GSM 4979; Shrewley, Warwickshire)

e *Pholadomya? richardsi*; ×2 (Holotype: GSM 4980; Shrewley, Warwickshire: possibly *Palaeocardita austriaca*)

f indeterminate bivalve, cf. *Lopha*; ×2 (L.9151; Shrewley, Warwickshire)

g, h *Acrodus keuperinus*; ×2.5 (g—GSM 90478, h—GSM 90478a; Shrewley, Warwickshire)

i *Acrodus keuperinus*; ×1 (GSM 4981; Shrewley, Warwickshire)

j *Dictyopyge superstes*; ×2 (Holotype: P 7614; Rowington, Warwickshire)

k Chirotherioid footprints; ×0.3 (G 1143, P B Brodie collection; Shrewley, Warwickshire)

Repository of specimens indicated by prefixes to registered numbers: BU—Birmingham University Museum; G—Warwickshire Museum; GSM—British Geological Survey, Keyworth; L, P—The Natural History Museum, London

Lilstock Formation

The Cotham Member consists of pale and medium grey, commonly blocky, calcareous mudstone with a few thin siltstone and limestone beds. The lowest 1.6 m proved in the Knowle Borehole are darker grey and siltier, and contain reworked fragments of Westbury Formation mudstone. A 13 cm limestone overlying the Cotham Member in the Knowle Borehole (Figure 11) may be an attenuated remnant of the Langport Member which is widely represented in the area to the south and east (Williams and Whittaker, 1974; Old et al., 1987). The paler, calcareous Cotham Member is commonly regarded as having formed in shallow, fresh or brackish waters. However, marine microplankton in these beds (Figure 13) are comparable with those of the Westbury Formation, indicating a continuing marine influence.

Palaeontology of the Penarth Group

A rich marine bivalve fauna in the Westbury Formation includes *Chlamys valoniensis*, *Eotrapezium concentricum*, *E*. cf. *germari*, *Lyriomyophoria postera*, *Modiolus* sp., *Protocardia rhaetica*, *Rhaetavicula contorta* and *Tutcheria* together with the gastropod '*Natica*' *oppeli* and fish such as *Gyrolepis alberti* (Figure 12).

The Cotham Member is largely devoid of macrofossils apart from *Euestheria minuta*. The possible Langport Member limestone at Knowle contains *Pteromya* cf. *tatei*.

Rich and varied palynomorph assemblages occur through the Penarth Group, in marked contrast to the underlying Mercia Mudstone. They have been recovered from Knowle Borehole (Figure 13), the Round Hill Trial Pit, Wootton Wawen (Figure 15), and from pits at Bentley Common [97 50], Mount Farm [9965 6310] and Upper Berrow Farm

a

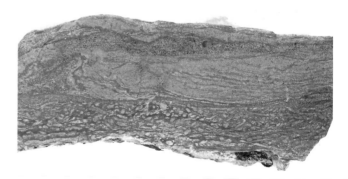

b

Plate 12 Bioturbation in the Arden Sandstone, Rowington, Warwickshire, ×1. (British Geological Survey: a—GSM 118453, plan view; b—GSM 118454, in section).

[9849 6302] in the south-west of the district. The assemblages comprise terrestrial miospores and marine organic-walled microplankton, and are indicative of a Rhaetian age. The massive and sudden increase in the abundance and variety of terrestrial miospores in the Penarth Group indicates rapid expansion and diversification of the land flora, probably in response to an amelioration of the climate and other conditions which may, in turn, have been associated with changes in the relative distribution of land and sea during latest Triassic times (Warrington, 1981).

Figure 11 Comparative sections in the Penarth Group

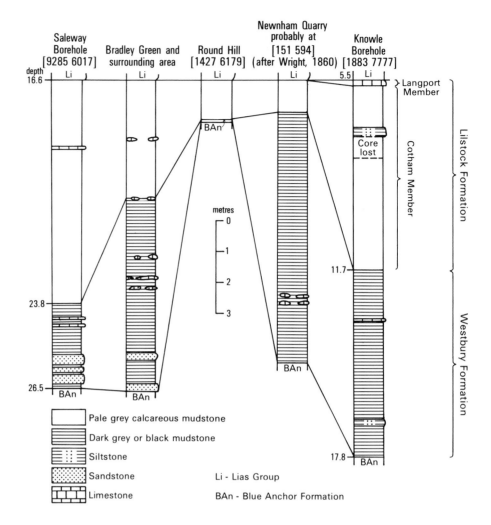

Figure 12   Upper Triassic and Lower Jurassic fossils

1   *Rhaetavicula contorta* (Portlock) × 1½ , Westbury Formation
2   *Lyriomyophoria postera* (Quenstedt) × 2,, Westbury Formation
3   *Euestheria minuta* var. *brodieana* (T R Jones) × 6, Lilstock Formation, Cotham Member
4   *Modiolus hillanus* J Sowerby × 1½ , Penarth Group to Blue Lias Formation
5   *Liostrea hisingeri* (Nilsson) × 1, Penarth Group to Blue Lias Formation
6   *Cardinia ovalis* (Stutchbury) × 1½ , Blue Lias Formation
7   *Astarte* sp. × 1½ , Blue Lias Formation
8   *Chlamys valoniensis* (Defrance) × 1½ , Penarth Group
9   *Plagiostoma giganteum* J Sowerby × ¾ , Blue Lias Formation
10   *Caloceras intermedium* (Portlock) × 1, Blue Lias Formation

Sources: 1, 2, 4, 5, 7, 8 and 9   *British Mesozoic fossils.* The Natural History Museum, London.
  3   Jones, T R. 1863.   *Monograph of the Palaeontographical Society,* pl. II, fig. 12.
  6   Stutchbury. 1842.   *Annals. Magazine of Natural History,* Vol. 8, pl. X.
10   Portlock, J E. 1843.   *Report on the Geology of the county of Londonderry etc.* (Dublin: HMSO.)

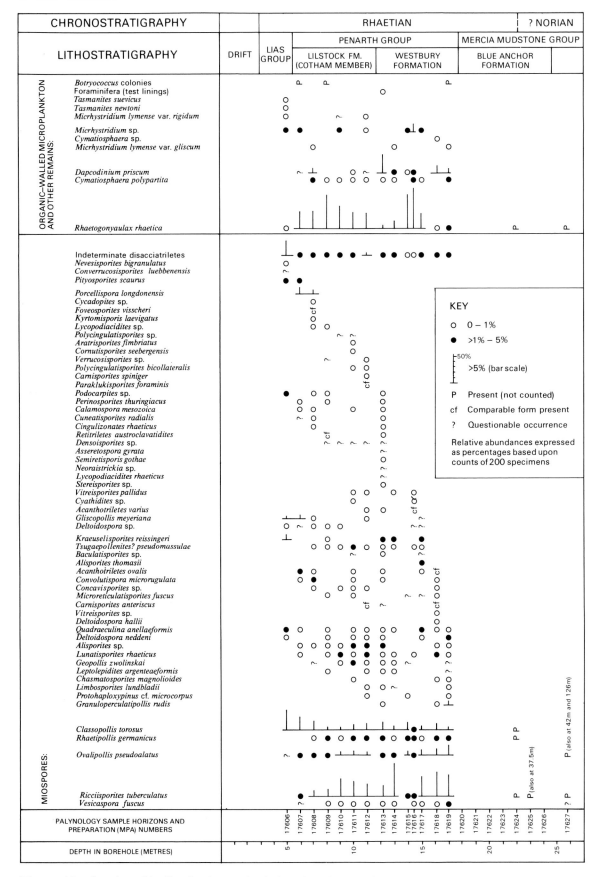

Figure 13 Stratigraphic distribution and relative abundances of palynomorphs from the Mercia Mudstone, Penarth and Lias groups of the Knowle Borehole

SEVEN

Jurassic

LIAS GROUP

Rocks of the Lias Group crop out in the south, between Aston Cantlow and Bearley, and in faulted outliers near Shelfield and Morton Bagot; in the south-west, around Bradley Green and Foster's Green; and in the north, at Copt Heath near Knowle (Figure 1). All the Lias strata present in this district belong to the Blue Lias Formation (Cope et al., 1980) which lies at the base of the Lias Group (sensu Powell, 1984). Beds below the lowest occurrence of the ammonite *Psiloceras planorbis* are of Triassic age, but are described in this chapter for convenience.

The sequence consists of alternating grey, calcareous mudstones and subordinate beds of limestone, and reaches a total thickness of about 75 m. It is divided into three members: at the base the Wilmcote Limestone Member, named here, has its type section at Wilmcote Quarry [151 594]; the overlying Saltford Shale Member was named by Donovan (1956) and has its type section in Saltford railway cutting near Bristol; the succeeding Rugby Limestone Member, named here, has its type section at Parkfield Road

Quarry, Rugby [SP 493 759], described by Clements (1977) and Old et al. (1987). In the west of the district, the upper part of the Wilmcote Limestone Member passes laterally into the Saltford Shale; the lower part of the latter in this area is of Triassic age (Figure 14; Table 3).

The rocks were laid down in a tropical shelf sea, with the nearest land, the slowly sinking London Platform, to the south-east. As the sea transgressed in that direction, successively younger beds encroached further onto the platform so that this district became progressively further offshore (Donovan et al., 1979). The proximity of the shoreline is shown by the remains of insects, terrestrial reptiles and plants found in the earliest sediments. Slight fluctuations in sea level gave areas of shallower water in which limestone was laid down, and shell-fragmental limestones were produced where current activity caused winnowing of finer sediments. Bituminous paper shale was deposited where conditions on the sea floor were stagnant and anaerobic. Although anaerobic conditions are commonly associated with a deep water environment, the Lias Group facies in general is indicative of deposition in shallow water in depths

Table 3 Lithostratigraphy of the Lias Group on Sheet 183, with comparative nomenclature of previous workers in this and neighbouring areas

Zone	Subzone	Lithostratigraphy	Lithology	George (1937)	Wright 1878	Wright 1860	Richardson (1904, 1905)	Others	Old et al. (1987)	Edmonds et al. (1965)
Arietites bucklandi —?—?—?	*Vermiceras conybeari* —?—?—	Rugby Limestone Member	Alternating mudstones & argillaceous limestones (cementstones) *up to 40 m*	Shales and limestones	Bucklandi Beds	Lima-Beds	angulata-Beds	Lima Beds (Brodie 1865, 1868, 1874, 1875)	Blue Lias	Blue Lias
Schlotheimia angulata	*Schlotheimia complanata*									
	Schlotheimia extranodosa					Buck-landi Beds				
Alsatites liasicus	*Alsatites laqueus*	Saltford Shale Member	Fissile & blocky mudstones with a few thin & nodular limestones *12–30 m*	Psiloceras shales	Planorbis Beds	Ammonites planorbis Beds	planorbis Beds		Lower Lias	Clays between White & Blue Lias
	Waehneroceras portlocki									
Psiloceras planorbis	*Caloceras johnstoni*							—?—		
	Psiloceras planorbis	Wilmcote Limestone Member	Alternating mudstones & limestones; shelly & argillaceous at base, laminated at top *1.3–10 m*	Pre-planorbis or Ostrea Beds	Ostrea or Saurian Beds	Ostrea Beds / O. liassica Beds / Ostrea Series / Saurian Beds	Ostrea Beds / Pre-planorbis Beds / Saurian Beds	Saurian Beds[2] / Ostrea Beds[1]	Insect and Saurian Beds[2]	

The term Lima Beds is not precisely defined, but appears to correspond to the Rugby Limestone Member.

The limits of the 'angulata' and 'Bucklandi' Beds are not defined.

Wright (1878) and Richardson (1904, 1905) restricted use of the term Saurian Beds to strata occurring below the *planorbis* Zone. Brodie (1868) appears to have included all the basal limestone–shale sequence

(Wilmcote Limestone Member) in his Saurian Beds and Insect and Saurian Beds. Brodie (1865) gave no clear definition of the top of the Saurian Beds, but he equated them with beds containing '*Ammonites planorbis*'.

The ammonite zones follow Dean et al. (1961).

1 Phillips 1871   2 Brodie 1868

below wave base. Thus the bituminous shales are thought to have been laid down on an extensive continental shelf of very shallow gradient in which minimal water circulation allowed stagnant conditions to develop (Hallam, 1967; Old et al., 1987).

Contemporaneous movement along the westerly-throwing Lickey End (Inkberrow) Fault bounding the Worcester Basin may also have influenced sedimentation patterns during the early part of Lias Group deposition. Movement in early Blue Lias times may have resulted in a localised area of deeper water west of the Fault, around Bradley Green, in which lime deposition was restricted.

In the l9th and early 20th centuries there were many working quarries in the district, mainly in the Wilmcote Limestone. From these sections, which contain diverse faunas, several classifications of the stratigraphical sequence were attempted (Table 3; Figure 14).

The base of the Jurassic system is marked biostratigraphically by the first appearance of ammonites of the genus *Psiloceras* (Cope et al., 1980). Since this fossil appears a few metres above the base of the Lias Group in this district, the beds underlying this horizon are assigned a Triassic age (Table 3); they are, however, described here with the remainder of the Lias Group.

A list of fossils from the Lias Group is given in Appendix 5, and selected fossils are illustrated in Figure 12.

Wilmcote Limestone Member

The Wilmcote Limestone Member consists of the interbedded limestones, mudstones and paper shales, including the Stock Green Limestone, at the base of the Lias Group. The type section at Wilmcote Quarry [154 594] (Plate 13; Figure 14, columns 8a and 8b) is no longer fully exposed; here, the base of the member is taken below the lowest limestone bed, the 'Guinea Bed' or Bed 29 of Wright (1860) (see Figure 14, column 8b) which rests on the underlying Cotham Member. The average thickness of the member is about 8 m near Wilmcote, thinning to about 5 m at Round Hill [1427 6179] but increasing in the north of the outlier to 8 to 10 m, and maintaining this thickness at Morton Bagot and at Knowle. The proportion of limestone is variable both vertically and laterally; in the Stock Green Limestone it is commonly 45 to 50 per cent, while in the overlying beds the proportion drops to 15 to 20 per cent.

Most of the beds in the Wilmcote Limestone appear to be a primary carbonate deposit; minor undulating interfaces and nodular horizons are the only features suggesting secondary or diagenetic migration of lime.

Correlation of individual limestones within the Wilmcote Limestone is uncertain. Most published quarry sections in the Binton–Wilmcote area have local names assigned to individual beds but the names are not consistent from quarry to quarry. Exceptions are the 'Guinea Bed' (see above), the 'Firestones' (a group of limestones just above the Guinea Bed), the 'Grizzle Bed' (a fossiliferous limestone marking the base of the *planorbis* Zone) and the 'Top Rock' (the highest limestone). Brodie (1874, 1875) reported finding the 'Guinea Bed' and 'Firestones' on the outliers north of Wilmcote (although they are absent at Round Hill), and they may occur above the 2.5 m of mudstone proved at the base of the Lias Group in the Knowle Borehole (Old, 1982; Figure 14, column 10). The 'Firestones' are equivalent to the Upper Stock Green Limestone of Ambrose (1984).

Brodie (1845) referred to the lowest one or two limestones of the Lias Group in Gloucestershire and Worcestershire as the 'Insect Beds or Limestones', and these are correlated by Worssam et al., (1989) with the lowest two limestones in the Twyning Borehole, near Tewkesbury (Figure 14, column 1). However, Brodie also correlated these two beds, on the basis of lithological and zoological similarity, with several insect-bearing limestones which span the whole of the Wilmcote Limestone in the Bidford, Binton and Wilmcote areas, despite the fact that the insect-bearing beds in south-west Warwickshire are clearly younger (Figure 14). Brodie (1874, 1875) reported finding the 'Insect Limestone' on the outliers at Round Hill, Morton Bagot and Knowle.

The limestones are generally fine grained, argillaceous and flaggy weathering, and in the upper part of the sequence most are laminated. They are commonly shelly or shell-detrital, and recrystallisation has affected some of the beds. The lowest limestone (the 'Guinea Bed') around Wilmcote and Binton is conglomeratic. Although most workers have placed it in the Lias, Tomes (in Woodward, 1893) placed it in the 'Rhaetic', and Woodward (1887) states that 'the White Lias' (now part of the Langport Member) 'is replaced by the Guinea Bed'. Arkell (1933, p.195) suggested that erosion in early Hettangian times locally removed the 'Langport Beds' in the Wilmcote area to form the 'Guinea Bed'. However, in view of the absence of Langport Member limestones (White Lias) west of Stratford-upon-Avon (Williams and Whittaker, 1974) it is more likely that the conglomeratic facies of the 'Guinea Bed' represents eroded shoreline debris transported from the east.

The Stock Green Limestone persists over the entire Lias outcrop in the south and west, and is readily identifiable in sections and in field debris. In the west of the district (Ambrose, 1984) it is the sole representative of the Wilmcote Limestone and is only about 1.3 m thick in the type area, (Figure 14, column 4), although George (1937) estimated up to 6.1 m for these ('pre-*planorbis*') beds.

Farther east it is identified only in quarry sections (Figure 14, columns 5 to 8b) where its position relative to higher, lithologically similar limestones is clear, but it did not prove possible to map it separately from the rest of the Wilmcote Limestone.

At Bentley Common, a trial pit exposed 0.9 m of weathered mudstone below the lowest beds of the Stock Green Limestone. The mudstone could not, with certainty, be assigned on lithological grounds to either the Lias Group or the Cotham Member but has provisionally been included with the former. Typical Lias Group mudstone was seen beneath the Stock Green Limestone in a temporary excavation at Broughton Hackett, 7.5 km south-west of Stock Green (Fig.14, column 2; Ambrose,1988). Taken with the palynomorph evidence at Bentley Common (see below), this suggests that Lias Group mudstone may occur at the base of the Stock Green Limestone over most or all of its outcrop in the south-west of the district. Elsewhere, mudstone occurs at the base of the Wilmcote Limestone at Round Hill (Strange and Ambrose, 1982) and Knowle (Old, 1982). The former is an attenuated sequence due to contemporaneous movement

Plate 13   Interbedded limestones and fissile mudstones in the Wilmcote Limestone, Wilmcote Quarry [150 593]. (A10835)

on the Vale of Morton Anticline (Chapter 9), and is probably all of *planorbis* Zone age, in contrast to the pre-*planorbis* age of the basal mudstones in the south-west of the district and at Knowle, which represent an earlier stage in the transgression of the Liassic sea. The Stock Green Limestone in the south-west is, in its lower part, fine grained and recrystallised; in thin section (E 57641) it is a finely laminated biomicrite or biomicrosparite composed of fine shell debris, remnant micrite pellets, silt grade quartz, scattered glauconite and opaque grains in a microsparitic matrix. The upper part consists of three or four beds of argillaceous, shelly limestones with mudstone partings; in thin sections (E57642–4) these limestones are seen to be composed of biosparite, biomicrosparite and biocalcarenite, with coarse clasts of bivalve and echinoid debris, in a micritic, microsparitic or coarse sparry calcite matrix.

The biostratigraphy of the Wilmcote Limestone is shown in Table 3 and Figure 14. Although Wright (1860, 1878) and Brodie (1868) cited the occurrence of an unspecified ammonite in the Stock Green Limestone at Binton, just south of the district, ammonite faunas collected at several other localities place the base of the *planorbis* Zone and Subzone at a higher level, such that the Stock Green Limestone is contained wholly within the Triassic. The specimen cited may have been assigned to the wrong bed in the Binton quarry section. The base of the *johnstoni* Subzone has been closely determined only at Wilmcote Quarry [151 594] (Figure 14, column 8a), where Dr D E Butler collected *Psiloceras* sp. and *P. plicatulum* from the *planorbis* Subzone and *Caloceras* sp., *C. intermedium ?*, *C. johnstoni* and *Psilophyllites?* from the *johnstoni* Subzone. At Round Hill [1427 6179], *Caloceras* sp. occurs only 3.2 m above the base of the Wilmcote Limestone, which in turn overlies a much attenuated Penarth Group and Blue Anchor Formation, indicating that localised uplift had occured on the Vale of Morton axis in Rhaetian and early Hettangian times (Figure 14, column 9). At Knowle, *Psiloceras planorbis* has been found in debris ploughed in the fields, but the detailed sequence hereabouts is not known.

In addition to ammonites, the Wilmcote Limestone yields plant remains, a rich and varied fauna of corals, abundant bivalves, (particularly *Liostrea hisingeri),* gastropods, crinoid and echinoid debris, ostracods, insects, and fish and reptile remains (Appendix 5).

The palynomorph assemblages from the Wilmcote Limestone in the Round Hill section (Figure 15) compare closely with those from the basal part of the Lias Group across the Midlands, although they also bear some resemblance to those commonly found in the Langport Member of the Penarth Group. A trial pit sited on the Stock Green Limestone at Bentley Common [9725 6570] was sampled from three levels down to 0.9 m below the lowest limestone. The very sparse palynomorph assemblages recovered (Appendix 5) are similar to those known from the basal (late Triassic) beds of the Lias Group, rather than those from the Lilstock Formation of the Penarth Group, and if the pit reached the Lilstock Formation it probably did not penetrate to the level of the Cotham Member.

Saltford Shale Member

The Saltford Shale consists of dark grey mudstone with subordinate paper shale and a few thin beds of limestone. It occurs in all the main Lias Group outcrops, varying in thickness from about 12 m at Round Hill to 20 m at Wilmcote and 30 m near Bradley Green. The westwards thickening is at the expense of the Wilmcote Limestone, and the facies change is attributable to increasing distance from the London Platform, and possibly to fault-controlled deposition associated with the Lickey End Fault. The mudstones are both fissile and blocky, fossiliferous and locally pyritic. In the west, fissile mudstone and paper shale are common in the lowest 6 m (Ambrose, 1984). The few limestones include shell-fragmental limestone with abundant *Cardinia* sp. near Broughton Green [957 610], laminated limestone near Wilmcote, and argillaceous rubbly limestone at Stock Green and Knowle.

The member ranges in age from Rhaetian to Hettangian (high *liasicus* Zone) in age. Beds of Rhaetian age are found only in the west (Figure 14, columns 2 to 4) where up to 2.6 m of mudstone underlie the lowest occurrence of *Psiloceras planorbis* (Ambrose, 1984). A thin limestone with abundant *P. planorbis*, marking the base of the Hettangian (base of *planorbis* Zone) at Stock Green, is correlated with the 'Grizzle Bed', a limestone within the Wilmcote Limestone, farther east. The higher part of the western sequence contains abundant *Cardinia hybrida*, *C. ovalis* and *C.* sp., suggesting a *liasicus* Zone age.

In the east, the base of the member is within the *johnstoni* Subzone. The exact biostratigraphical horizon of the top of the member is not precisely known, but it probably varies from place to place, and may locally extend into the overlying *angulata* Zone. Old et al. (1987) have demonstrated a similar variation in the area between Harbury and Rugby to the east. At Knowle there is an estimated 25 m of the member and the occurrence of *Cardinia ovalis* suggests that the *liasicus* Zone may be present (Old, 1982).

The fauna of the Saltford Shale is poorly known owing to the lack of boreholes and sections. Bivalves, particularly *Cardinia* sp., are common; a coral, ostracods and crinoids were also found, together with burrows, including *Chondrites*, in the paper shales.

Rugby Limestone Member

The Rugby Limestone Member consists of alternating grey, argillaceous limestones and dark grey mudstones, and is synonymous with the 'Blue Lias' of the Rugby–Harbury area (Old et al., 1987). The base is easily recognised in the field by the incoming of abundant pale grey rubbly limestone debris, which is distinct from the flaggy weathering limestones which typify the Wilmcote Limestone. The western outcrops near Stock Green are part of a larger outcrop extending southwards to Gloucestershire. There are outliers at Broughton Green [957 606], Berrow Hill [997 623], and at Wilmcote and Round Hill.

The full thickness of the member is nowhere preserved; a maximum of 20 m is present at Wilmcote and Berrow Hill and at least 40 m in the south-west (Ambrose 1984). At Round Hill about 10 m are preserved, and at Broughton Green only the lowest 2 to 3 m crop out. There are no exposures in this district and the strata are known only from debris in fields and ditches.

The Rugby Limestone in this district ranges from high *liasicus* Zone to *angulata* Zone in age and may, as in neighbouring areas, extend into the *bucklandi* Zone. The former two zones were proved in the south-west by the presence, respectively, of *Waehneroceras* spp. and *Schlotheimia* spp. The non-ammonite fauna includes foraminifera, bivalves and ostracods.

The limestones of the member are similar to those in the Rugby area, for which a composite primary and secondary origin, proposed by Hallam (1964), is favoured by Old et al. (1987).

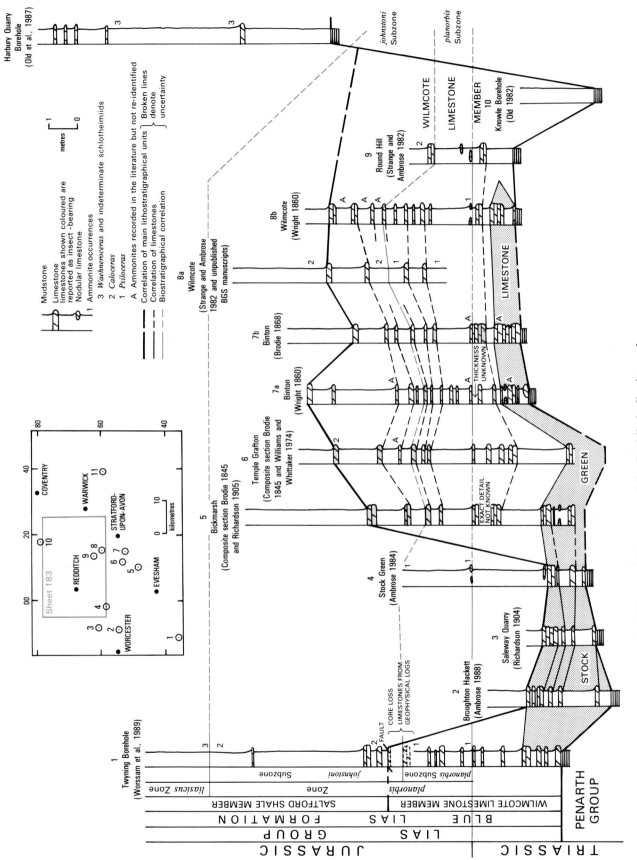

Figure 14 Comparative sections in the Wilmcote Limestone of the Redditch district and adjoining areas

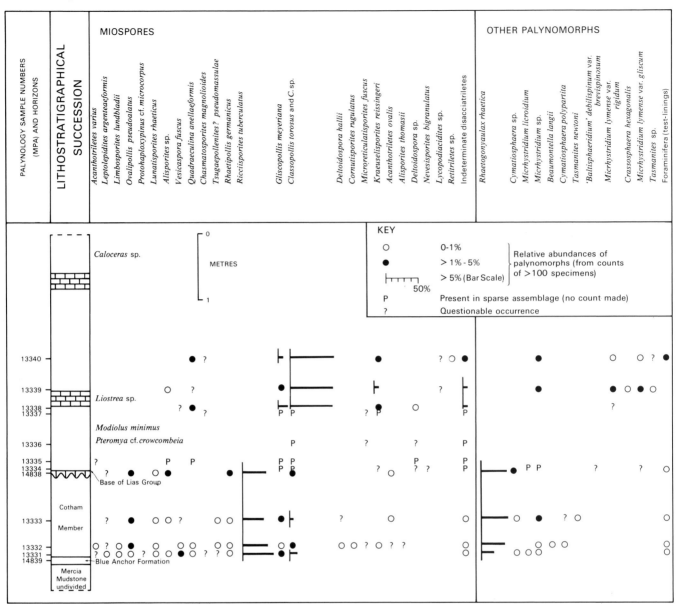

Figure 15   Distribution of molluscs and palynomorphs in Round Hill trial pit, Wootton Wawen
[1427 6179]

EIGHT

Quaternary

GLACIAL DEPOSITS

All the glacial deposits are the result of one period of glaciation. In the east, they include lithologies of eastern provenance which correlate with those of the type Wolstonian
deposits between Coventry, Rugby and Leamington Spa
(Shotton, 1953). The nomenclature and correlation of the
deposits follows that of Sumbler (1983), which in turn is
based on Rice (1968) and Shotton (1976). In the present
district, however, the flat-bedded, widely persistent outcrops
of the type area are not well developed, in part due to the
uneven subdrift topography (Figure 16). Instead, the boundaries of the lithological units are irregular and the units tend
to interdigitate. In particular, there is an intermingling of
the Triassic-derived and chalky boulder clays of eastern
provenance which, in the Warwick district, usually crop out
separately (Old et al., 1987). The glacial deposits in the west
are of western provenance. In places, they interdigitate with
the youngest eastern deposits and are presumed to be contemporaneous with them (Table 4).

The status of the Wolstonian glacial stage is uncertain,
and has been the cause of much debate. Shotton (1989)
maintained his earlier view (Shotton and West, 1969) that
the type Wolstonian deposits postdate the Anglian Stage.
However, Sumbler (1983) tentatively regarded them as
Anglian in age, a view supported by Old et al. (1987), Rose
(1989) and by the present writers. Nevertheless, contemporaneous fossils are rare in this district and, in particular, no
interglacial faunas or floras like those of the Birmingham
area (Kelly, 1964; Horton, 1974) have been found. The
term Wolstonian is retained in this account pending an
assured assignation to the Anglian.

EASTERN GLACIAL DRIFT

Baginton Sand and Gravel

Sand and gravel around Snitterfield [21 59] contains up to 32
per cent of Jurassic limestone pebbles and derived fossils
(Cannell and Crofts, 1984) but no chalk or flint. The
deposits reach 5 to 6 m in thickness and commonly consist of
sand, up to 2 m thick, overlying sand and gravel. The base of
the deposits ranges from 84 to 92 m above OD and falls gently to the south-east and north-east. A recently excavated
temporary pit [235 597] showed the top of the sand inter-
digitating with the overlying Wolston Clay (Rose, 1987).
This pit yielded an almost complete skull of *Palaeoloxodon antiquus* together with aquatic and terrestrial molluscs (Lister
and Keen, 1989). At the former Hutchins Brickyard, Snitterfield [2375 5932], the gravels yielded a molar of either
Palaeoloxodon antiquus (Lucy 1872; Tomlinson 1935, p.436) or
Mammuthus primigenius (Lister and Keen, 1989). Shotton

(1953, pp.215, 240–241) correlated these deposits with the
Lillington (Jurassic-bearing gravels) facies of the Baginton
Sand and Gravel, and the Jurassic clasts are probably derived from outcrops to the south and west. Patches of sand
and gravel at a similar level to the north, near Norton Lindsey, lack the Jurassic component and may have been derived
contemporaneously from the north.

Red Triassic-derived boulder clay (Thrussington Till)

Boulder clay with a red-brown matrix derived from the Mercia Mudstone characterises many of the outcrops east of the
Kingswood Gap and north of Wolverton. In places, it occurs
at the base of the glacial sequence, but elsewhere it lies between lake deposits and sand and gravel. It forms much of
the boulder clay plateau east of the Kingswood Gap, where
commonly it is closely associated with flinty and chalky
boulder clay. In boreholes, it is usual for chalky boulder clay
to overlie non-chalky boulder clay (Cannell and Crofts,
1984). Erratics include well-rounded pebbles of quartzite or
quartz, presumably derived from the Kidderminster Formation, Triassic sandstone and a few small coal fragments;
chalk and flint are absent. The lithology of the till closely
resembles that of the Thrussington Till of the Warwick
district (Sumbler, 1983). In the present district, however, a
few boreholes have proved glacial lake deposits and sand and
gravel within the boulder clay (Cannell, 1982, pp.43, 48), in
contrast to the situation further east where the Thrussington
Till forms a single bed overlying the Baginton Sand and
Gravel (Rice, 1968; Shotton, 1976).

The flint and chalk-free boulder clay of the Wroxall area
was included by Tomlinson (1935, fig. 1) with the 'Western
Drift'. Shotton (1968a, p.58), however, found no western
erratics in the boulder clay between Oldwich House and
Frogmore Wood, and concluded that it was of northerly
derivation. During the present survey, many angular, grey,
quartzite erratics were found between Clattyland Wood and
Honiley. These are comparable with both the Lickey and
Hartshill quartzites, which crop out to the west and north-
east respectively, and so do not show an unequivocal direction of derivation of the till.

The boulder clay varies rapidly in thickness, particularly
where it fills subdrift valleys. Exceptional thicknesses of up to
18 m were proved in boreholes at Norton Lindsey [2215
6381] and Hay Wood [2063 7082], but boreholes nearby
proved only a few metres of boulder clay (Cannell and
Crofts, 1984).

Glacial lake deposits (Wolston Clay)

The glacial lake deposits are mainly red-brown, laminated
clays and silts with minor beds of sand. Laminations are not
strongly developed everywhere and some of the deposits

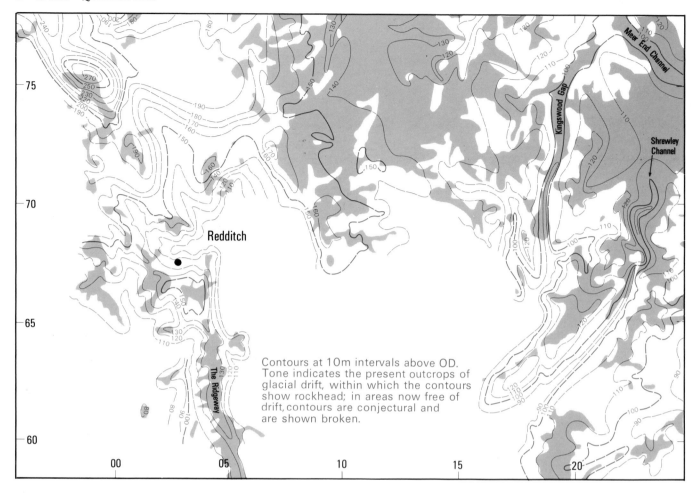

Figure 16   Contours on the base of the glacial drift

grade into red, Triassic-derived till. The deposits are commonly interbedded with thicker accumulations of sand and gravel and boulder clay; thus all show rapid variations in thickness, which are compounded by the fact that they were were laid down upon an uneven subdrift topography. They are restricted largely to the area east of the Kingswood Gap; within the Gap itself, deposits are rare. Their maximum

recorded thickness of 15.6 m, which included 2.6 m of sand and gravel, was proved in a borehole near Frogmore Farm [2363 7535] (Cannell, 1982, p.46).

The lithology of the lake clays suggests a correlation with the Wolston Clay of the Warwick district (Shotton, 1953; Sumbler, 1983). This correlation has been made for the laminated clays exposed at Hutchins Brickyard, Snitterfield,

Table 4   Stratigraphy of the Drift deposits

FLANDRIAN	Alluvium, Peat, Landslip		
DEVENSIAN	First Terraces, Lacustrine Alluvium Avon Second Terrace, Severn Third Terrace, Head		Periglacial Flood Gravels and Alluvial Fans
IPSWICHIAN	Avon Third Terrace		
'WOLSTONIAN'	Avon and Severn Fourth Terrace	WESTERN GLACIAL DRIFT Boulder Clay Sand and Gravel Earlswood Till Glacial Lake Deposits	EASTERN GLACIAL DRIFT Chalky and Flinty Boulder Clay (Oadby Till) Sand and Gravel (Wolston Sand and Gravel) Glacial Lake Deposits (Wolston Clay) Triassic-derived Boulder Clay (Thrussington Till) Baginton Sand and Gravel

(Shotton 1953, pp.240–241) and in another pit nearby (Rose, 1987) where they overlie the Baginton Sand and Gravel. Elsewhere in the present district, the lake deposits are interbedded and interdigitate with other glacial lithologies, and the absence of the Baginton Sand and Gravel makes the correlation less clear. The top of the glacial lake deposits ranges from 101 m above OD in the south to 126 m above OD between Honiley and Balsall Common. The height range is consistent with that of the Wolston Clay (Shotton, 1953, p.247), and distinct from that of the lake deposits of western origin which lie at a higher level, as described below.

Sand and gravel (Wolston Sand and Gravel)

Sand and gravel is interbedded with the other glacial deposits east of the Kingswood Gap and forms the bulk of the deposits within the Gap itself. It occurs at all levels within the sequence, but mostly predates the youngest boulder clay of the Wroxall plateau. In the north, the clasts are almost exclusively of quartzite and quartz, with rare sandstone and mudstone. South and south-west of Shrewley, flint and chalk pebbles appear in increasing numbers and are abundant at Tattle Bank [183 639] (Tomlinson, 1935). A few flints also occur in the gravels around Heronfield [19 74] and in boreholes as far west as Lapworth (Cannell and Crofts, 1984), but the predominant pebble lithologies west of the Kingswood Gap indicate a western origin.

If the glacial lake deposits are equivalent to the Wolston Clay, then the sand and gravel interbedded with them may correlate with the Wolston Sand of the Warwick district (Shotton, 1953), although the sequence in the present district is more complex and more closely resembles that of the Rugby area (Sumbler, 1983).

As the glacial deposits are traced westwards towards the Kingswood Gap, sand and gravel becomes dominant in the sequence, emerging from beneath till along both sides of the Gap; it is contiguous with, and therefore presumably contemporaneous with deposits laid down within the Gap. At 'Riley's pit', Cuttle Pool [2015 7535] (Tomlinson, 1935, p.429; Shotton, 1985), small exposures in cross-bedded sand showed southerly dipping foresets, which indicate that the deposits were laid down by melt water flowing south through the Kingswood Gap. The sand and gravel deposits generally vary rapidly in thickness, especially where they fill channels. An unusual thickness of 31.1 m was proved in a borehole at Haseley [2321 6793].

Chalky and flinty boulder clay (Oadby Till)

Till, with many flint and chalk clasts, and a few of Jurassic origin, becomes increasingly common southwards along the eastern side of the district. The clay matrix is typically brown and rarely grey. This chalky boulder clay is younger than the lake deposits and most of the sand and gravel. Usually, it is only a few metres thick, but 14 m were proved in a borehole at Bearley [1926 6064]. In the north, it is contiguous with the upper part of the Triassic-derived boulder clay and has not been separately mapped thereabouts. Boreholes proving chalky boulder clay commonly show an underlying red, non-chalky till (Cannell, 1982; Cannell and Crofts, 1984), and in a borehole near Wroxall [2152 7154] a 0.7 m layer of chalky boulder clay lay within red, non-chalky till (Cannell and Crofts, 1984). South and west of Wolverton, the boulder clay is entirely flinty and, in part, chalky, and Wills (1937, p.81) records eastern erratics along The Ridgeway. The erratic suite of this boulder clay is comparable with the Oadby Till (Rice, 1968; Sumbler, 1983), although the latter has a grey rather than a brown clay matrix.

WESTERN GLACIAL DRIFT

To the west of the Kingswood Gap and The Ridgeway, the glacial drift is predominantly of western provenance.

Glacial lake deposits

The glacial lake deposits have their principal outcrop around Cheswick Green [13 75] where they fill a valley cut in Mercia Mudstone (Figure 16). The lake deposits are, in some places, the oldest in the glacial sequence, and elsewhere are underlain by sand and gravel. They are overstepped by younger glacial drift and, to the west and south, pass gradually into the complicated till deposits around Earlswood. Lake deposits up to 8 m thick have been proved in boreholes and at outcrop.

In the rare exposures of these deposits, they are seen to comprise red-brown, laminated clays and silts, probably derived mainly from local Mercia Mudstone, with varying amounts of sand. The lithology is identical to that of the lake deposits of eastern origin, but the height of the upper surface of the western lake deposits, ranging from 130 to 152 m above OD, does not coincide with that of their eastern counterparts. These deposits are probably the correlatives of those laminated and stoneless clays formerly exposed at California, Birmingham (Eastwood et al., 1925, pp.113–114; Pickering, 1956, p.232).

Earlswood Till

The Earlswood Till crops out around Earlswood and Trueman's Heath (see inset diagram on 1:50 000 sheet) and has been proved by boring beneath boulder clay in the same vicinity. It consists of up to 9 m of an intimate association of red-brown boulder clay, laminated clay, silt, sand and gravel. There is every gradation between these lithologies, and the thicker beds of each have been mapped separately.

The boulder clay is a stiff, red-brown clay, derived from the Mercia Mudstone and containing many small Triassic clasts of local origin and small amounts of coal and quartzite. It is lithologically distinct from the younger, orange-brown boulder clay described below. The laminated clays are similar to those described above but are interbedded with a variety of structureless clays, some pebbly, and silts and sands. In addition to those of the predominant red-brown colour, there are orange, grey and brown beds.

Sand and gravel

Sand and gravel occurs in close association with the lacustrine deposits and the Earlswood Till; like them, it is thickest in the subdrift valleys, thinning rapidly onto the interfluves. It is also commonly found underlying the younger boulder clay. Along The Ridgeway, many small exposures of boulder clay show thin bands of sand and gravel, interbedded with the till in a complex way. A maximum thickness of 18.8 m of sand and gravel is recorded in a channel at Money Lane [70 95] (Barton, 1960), and more than 12 m of sand and gravel occur at Cobley Hill, Rowney Green, Lowans Hill and Trueman's Heath. Such thicknesses do not persist over wide areas, however, and 5 to 10 m are more common. At many localities in the east, pebbles of 'Bunter' quartzite or quartz occur almost to the exclusion of other lithologies. In some outcrops, however , notably Tattle Bank, Welsh igneous and pyroclastic rocks occur (Tomlinson, 1935). Farther west, there is a greater variety of pebbles, including igneous lithologies of Welsh and Uriconian origin, chert, ironstone and sandstones of Carboniferous type. Exceptionally, pebbles of slate, phyllite and other metamorphic and igneous rocks predominate in pits up to 2.6 m deep at Hewell Farm [0020 6967].

Boulder clay

Most of the drift-covered plateau south of Hollywood and Cheswick Green is covered by boulder clay, generally 8 to 11 m thick. It is an orange-brown or red-brown clay, commonly with streaks and blotches of pale grey, and is variably sandy and pebbly. The erratics are almost exclusively quartzite and quartz pebbles, with rare flint in the south and, in the north, rare angular quartzite, comparable to the Lickey Quartzite, together with Triassic and Carboniferous sandstones. Unlike the boulder clay of the plateau east of the Kingswood Gap, there is no chalky facies. Boulder clay of the same lithology was proved in a borehole near Claverdon [1060 6686]. Normally the boulder clay overlies older drift and there is a topographical feature marking the junction. Between Whitlock's End [10 76] and Light Hall Farm [12 77], and around Bentley Heath [16 76], however, there is commonly no topographical expression of the contact and the relationship between the boulder clay and sand and gravel is not clear.

Glacial history

The subdrift (mainly preglacial) topography for the part of the district with the greatest drift cover is shown in Figure 16. The preglacial topography was modified by glacial and postglacial erosion, and thus the position of the preglacial watershed is, in part, conjectural. In particular, the channels at the Kingswood Gap, Meer End and Shrewley are probably at least partly glacial in origin. In the north, the preglacial drainage was northwards to the River Tame and in the south it was mainly southwards into the River Avon. In the extreme west, the levels of patchy outcrops of glacial drift indicate preglacial drainage directed westwards to the River Severn. In the south-east the preglacial drainage differed from that of the present day and flowed into the north-easterly 'proto-Soar' (Shotton, 1953).

The oldest Quaternary deposits are those of the Baginton Sand and Gravel around Snitterfield, which were laid down by the 'proto-Soar' river. Fossils from these deposits suggest temperate conditions (Lister and Keen, 1989) while those from farther north-east indicate a temperate climate followed by a cold, though not arctic one (Shotton, 1953, 1968b; Osborne and Shotton, 1968; Kelly, 1968; Gibbard and Peglar, 1989). At about the same time as these deposits were forming, ice advancing from the north or west blocked the Tame valley near Tamworth, leading to the formation of a lake extending into the Blythe valley in the present district. The surface of the lake reached over 100 m above OD, and its water probably overflowed south-eastwards through the Meer End Channel (Figure 16), into Finham Brook, enlarging it in the process (Tomlinson, 1935, p.449). The glacial deposits in and adjacent to the Meer End Channel relate mainly, however, to the Eastern Drift.

Ice advancing from the east blocked the outlet of the Meer End Channel, diverting the overflow of the lake firstly through the Wroxall Col at about 110 m above OD into the Shrewley Channel. The latter may be largely glacial in origin. As the eastern ice advanced, the Shrewley Channel also became blocked and the lake overflowed through the Kingswood Gap, the southern part of which is again largely of glacial origin. The probable south-westerly continuations of the Shrewley Channel and the Kingswood Gap (Figure 16) correspond with the courses of the Pinley Brook and the River Alne respectively. The Kingswood Gap channel was probably in operation longer than the previous two. Eventually it drained the lake which had by then become largely filled with sediments. Appreciable amounts of sand and gravel may have been laid down in the Kingswood Gap and in the deeper ice-free northern ends of the Meer End and Chadwick End valleys. Whether ice advancing from the east crossed the Kingswood Gap is not certain. Certainly, most of the drift to the west of it is of western type and flints are rarely found.

Next, the valley of the Avon was blocked by Welsh ice at Bredon Hill, near Evesham, and a lake (Lake Harrison), or series of lakes, was created in the Avon and 'proto-Soar' basins (Shotton, 1953; Sumbler, 1983). The eastern ice waned at about this time, depositing the earlier Triassic-derived boulder clay and, around the margins of the ice, glacial lake deposits and sand and gravel were laid down.

The eastern ice readvanced to reach the maximum limits shown in the inset diagram on the 1:50 000 map. Simultaneously, ice advanced from the west, blocking the streams draining northwards into the Tame and creating a lake or series of lakes from Hollywood to Dorridge. The Earlswood Till sequence was deposited close to the front of the Western Ice as it advanced into the lacustrine environment, while farther from the ice-front, large bodies, more clearly differentiated into glacial lake deposits and sand and gravel, were laid down. At their maximum extent, the two ice sheets probably impinged along the line shown in the inset diagram on the 1:50 000 map (cf. Wills, 1937, fig. 5), and the subsequent wasting of the ice left behind the great sheet of chalky, flinty and Triassic boulder clay in the east, and the orange-brown boulder clay in the west. The paucity of lake deposits and boulder clay in the Kingswood Gap may indicate that it was already partly blocked by sand and gravel before the flooding of the Avon valley. But it is also possible that meltwater flowing along the contact between the ice sheets eroded any such sediments. In the latter case, the sand and gravel of the Kingswood Gap would largely be outwash from the melting ice sheets.

POSTGLACIAL DEPOSITS

No glacial deposits of post-Wolstonian age occur in the district because the Devensian ice-sheet did not extend into the area. However, considerable erosion of the Wolstonian deposits and the underlying solid rocks occurred at this time, especially in the Avon basin, where post-Wolstonian erosion along the River Arrow exceeds 50 m. This erosion probably took place mainly under periglacial conditions when the Devensian ice front approached from the north to within about 15 km of the present district (Institute of Geological Sciences, 1977).

The postglacial deposits are mainly derived from the glacial drifts, with varying amounts also derived from the solid formations. In the absence of any fossils in this district the suggested chronology for these deposits (Table 4) relies largely upon correlation with fossiliferous deposits in adjacent areas. Tomlinson (1925, 1929, 1935) and other writers who followed her interpretation (e.g. Wills, 1938; Shotton, 1953), believed that the Avon Third Terrace deposits, of Ipswichian age, predated those of the Avon Fourth Terrace. Despite the lack of conclusive evidence, the authors of the Geological Survey memoirs dealing with the Avon basin (Edmonds et al., 1965; Williams and Whittaker, 1974; Old et al., 1987) adopted the contrary view that all the terrace-flats were aggradational, and thus that increased height of deposits above the alluvium equates with increased age. Recently, mapping and palaeontological work in the area to the south of the present district has confirmed that the deposits of the Fourth Terrace are older, and date from an undefined interglacial between the Hoxnian and Ipswichian stages, while the Third Terrace deposits are, indeed, of Ipswichian age (Bridgeland, Keen and Maddy, 1989; Worssam et al., 1989).

Periglacial flood gravels and alluvial fans

These deposits occur most commonly in the west of the district where post-Wolstonian erosion has been greatest. They consist mainly of sloping fans of bedded, clayey sand and gravel, but include all gradations between unsorted pebbly clay and clean gravel. In the area north-west of Redditch, they occur at three levels of decreasing height and age. The deposits of the highest two levels are much-dissected plateau remnants; the lowest deposits occupy floors and lower slopes of valleys, and grade into the First Terrace of the River Salwarpe.

The deposits formed by sheetflood, streamflood and braided fluvial processes in a periglacial environment; the unsorted and clayey beds may have been mudflows. The gravels are better sorted and coarser in grain-size than glacial sand and gravel and occur at lower levels, more closely related to the present topography. Unlike river terraces, the deposits slope appreciably into valley bottoms and lack the sharp, up-slope, marginal depositional features of the former. The flood gravels differ from head in being better sorted and bedded, and in having near-planar tops. Thicknesses of up to 5 m have been proved in boreholes, but generally the deposits are 1 to 3 m thick.

River terraces

River terraces are most common along the tributaries of the Avon and Severn, where there has been the greatest amount of postglacial erosion. The numbering of the terraces on these tributaries corresponds with the schemes of Tomlinson (1925, 1935) for the Avon, and Wills (1938) for the Severn. Profiles of the terraces of the River Arrow are shown in Figure 17. On streams draining into the Tame/Trent basin,

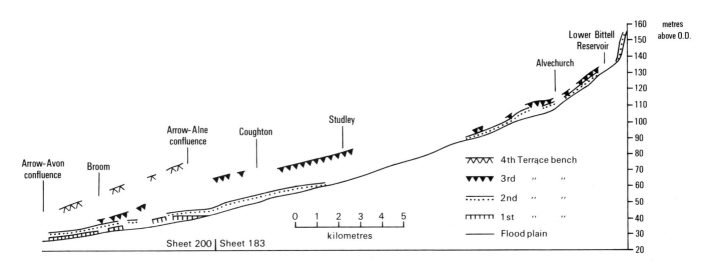

Figure 17 Long profile of the River Arrow and its terraces

and on a few draining into the Avon basin, a wider classification has not been attempted, and the numbering of the terraces is of only local significance.

FOURTH TERRACE

Small patches of the Avon Fourth Terrace occur in the south-eastern corner of the district. Much larger outcrops in the adjoining districts are described by Edmonds et al. (1965) and Williams and Whittaker (1974). In this district the deposits form gently sloping spreads about 1 m thick, 23 to 33 m above the alluvium, and consist of quartz and quartzite pebble gravel with a few flints. A single outcrop of Severn Fourth Terrace occurs in Bromsgrove [964 708] where T N George's field map records 1.5 m of sand and pebbles at Stratford Road. This locality is about 15 m above the alluvium of Spadesbourne Brook, and lies on the Fourth Terrace profile shown by Wills (1938, fig. 13).

THIRD TERRACE

Deposits of the Avon Third Terrace occur on the west side of the Arrow valley between Studley [07 64] and King's Coughton [08 58]. The terrace surface slopes gently and lies at 15 to 20 m above the alluvium. The deposits are estimated to be up to 3 m in thickness and consist of cobble gravel in a matrix of sandy clay, together with some thin sandy or clayey lenses. The cobbles are mainly quartzites with minor proportions of vein quartz and schorl. Deposits of the Severn Third Terrace lie along the Battlefield and Spadesbourne brooks, as shown by Wills (1938, fig. 13). They consist largely of Trias-derived, gravelly, clayey sands up to 4.3 m thick. Boreholes at Shenstone Teachers Training College [966 714] encountered up to 3.1 m of terrace deposits, mainly grey, sandy silt and dark brown, peaty silt overlying clayey sand and gravel.

SECOND TERRACE

The Avon Second Terrace has numerous outcrops along the Arrow and the Alne, where it lies typically 2 to 5 m above the alluvium, but on the east bank of the Arrow at King's Coughton [087 587] the terrace slopes gradually down to the alluvium. The deposits have rarely been proved to their full thickness but probably do not generally exceed 4 m. They consist mainly of quartzite gravel and clayey sand; flints are found in the deposits along the Alne. The outcrop at Wootton Pool [15 64] occupies a dry valley through which the River Alne probably had its former course.

Second Terrace deposits in the south-west corner of the district are tentatively correlated with those of the Avon preserved along the lower reaches of Bow Brook in the Worcester district.

FIRST TERRACE

First Terrace deposits occur along the River Blythe at Widney Manor [15 77] and east of Knowle [19 76]; along the River Rea at Longbridge [00 77]; at Haseley [23 68] and along Bow Brook and its tributaries in the south-west. The numbering of these terraces is of local significance only. They lie 1 to 2 m above the adjacent flood plain, and their deposits, generally less than 3 m thick, consist of gravel, sand, silt and clay in greatly varying proportions.

The Avon First Terrace grades into the alluvium of the Arrow and Alne at their confluence just south of the district (Figure 17), and is not developed farther upstream on these rivers.

Alluvium

The larger streams have more or less continuous alluvial flood plains, although these are commonly quite narrow. Wider areas of alluvium, such as at the headwaters of Cuttle Brook [190 755] and on the River Arrow at Studley [08 63], have apparently been augmented by colluvial downwash.

The alluvium typically comprises an upper layer of grey or brown clay and silt up to 3 m thick, underlain by several metres of bedded sand and gravel. Considerable variation is found even in the same river valley, however, and it is not uncommon for the whole deposit to be silt or clay.

Lacustrine alluvium

Along Brandon Brook near Feckenham [03 61], alluvial deposits at about the level of the local First Terrace are probably of lacustrine origin. The deposits consist of dark brown, grey and blue-grey clays with thin gravels. A gravel deposit, sinusoidal in plan, about 1 m thick and 75 m wide extends for 600 m near Shurnock Court [0244 6085 to 0188 6063]. Peat overlying the alluvium may represent the final filling of the lake with sediment.

Peat

Peat has formed in a few valleys where impeded drainage has resulted in the rapid accumulation of decaying vegetable matter. The best-documented deposit forms a bog 500 m north-east of Ipsley Alders Farm [074 673] and is probably the shrunken relic of a wider tract formerly covering the adjacent fan gravels. Boreholes have proved peat and organic clays to depths of up to 2.4 m. Samples of peat from an auger hole [0784 6755] yielded radiocarbon ages of 5430 ± 155 years BP and 6350 ± 155 years BP, at depths of 0.5 to 0.7 m and 1.2 to 1.5 m respectively (Welin et al., 1975). These dates place the formation of the peat within the postglacial Atlantic period (7450 to 4450 years BP), a time of warm, moist climate.

Two outcrops of peat, each more than 1.3 m thick, overlie the lacustrine alluvium of Brandon Brook. Peat, up to 1 m thick, overlies the alluvium for some 800 m to the north-east of Warren Farm [217 728]. The ground is waterlogged here because strong springs emerge at the base of the local sand and gravel deposits where they rest upon glacial lake deposits.

Head

Head, formed by periglacial solifluxion, is widespread, and has probably formed whenever conditions were favourable since the end of the Wolstonian. Only the thicker and more clearly defined deposits are shown on the 1:50 000 geological map. Gravelly wash occurs on most slopes underlain by or adjacent to outcrops of glacial drift or rocks of the Kidderminster Formation. The deposits vary according to the local source rocks, but are typically red or brown pebbly, silty clays, rarely more than 2 m thick.

Landslip

Landslips surround Berrow Hill [998 624], and involve the Rugby Limestone and the underlying Saltford Shale. The slip features are best preserved on the eastern slopes of Berrow Hill, where a prominent back scar has uneven ground below it with many small, fresh, slip scars.

NINE

Structure

The Redditch district can be divided into three principal structural areas, each with a northerly or north-north-westerly trend. From west to east these are the Worcester Basin, the Lickey Ridge and the Knowle Basin including part of the Warwickshire Coalfield. The Coventry Horst, to the north-east of the Warwick Fault (Figure 18), just impinges on the district. More details of the structure of the Lickey Hills area are shown on the 1:25 000 inset map on the 1:50 000 sheet, and of the main Sherwood Sandstone outcrop in Figure 3. An interpretation of part of the BGS Stratford 1986 seismic reflection profile (Chadwick and Smith, 1988) which relates to the district is shown in Figure 19.

The Worcester Basin and the subsidiary Knowle Basin form part of a wide post-Carboniferous pull-apart basin which, together with the Cheshire Basin to the north (Earp and Taylor, 1986), provided the main northward route for Permian and Triassic arenaceous sediments into the East Irish Sea Basin. The Worcester, Cheshire and East Irish Sea basins subsided at intervals during the deposition of the earlier Trias, and more or less continuously during the deposition of most of the Mercia Mudstone Group (Wills, 1948; Audley-Charles, 1970). In the light of recent research it seems likely that this whole pull-apart system, from the Worcester Basin northwards, was part of the tectonic cycle that culminated in the opening of the North Atlantic. Such a rift system can be viewed as a 'failed arm' of the North Atlantic and its progressive development controlled sedimentation throughout Permian and Triassic times (Kent, 1975; Owen, 1976; Whittaker, 1985).

The results of regional gravity and magnetic surveys which include the Redditch district are published at the 1:250 000 scale (see list of published maps, p.iv). The Bouguer gravity and aeromagnetic anomaly inset maps accompanying the 1:50 000 geological map have been extracted from these publications. As most of the geophysical anomalies reflect the effects of structures in both the 'basement' and 'cover' rocks, their interpretation is included in this chapter. Although the Bouguer gravity and aeromagnetic anomalies do not allow for a unique interpretation, they do provide additional controls on the deeper structure and augment the limited borehole data.

The gravity interpretation recognises that there are marked density contrasts at the bases of the Triassic and Upper Carboniferous successions, the older formations being generally denser, and that there may be other density contrasts in the largely unproven Lower Palaeozoic and Precambrian basement. All of these differences will contribute to the anomaly pattern as the thickness and disposition of the units varies.

The magnetic anomalies are unlikely to originate in the sedimentary sequences, which are not conspicuously magnetic. The anomalies more plausibly mark magnetic basement rocks or buried igneous intrusions, such as occur within the basement exposed in the Malvern Hills (Brooks,

1968), Shropshire (Greig et al., 1968) and Charnwood Forest (Worssam and Old, 1988).

Structures affecting the Mesozoic rocks, dominated by faulting associated with the Worcester Basin and the subsidiary Knowle Basin (Wills, 1956), were probably active throughout most of Mesozoic time. The faults have predominantly northerly or north-westerly trends; most have westerly downthrows, although some of those on the eastern side of the Lickey Ridge are important exceptions.

WORCESTER BASIN

The Worcester Basin is a major Triassic basin, floored by Precambrian and Lower Palaeozoic rocks, underlying much of the west Midlands (British Geological Survey, 1985; Whittaker, 1985; Chadwick and Smith, 1988; Worssam et al., 1989). In this district its eastern flank is marked by the Lickey End and Weethley faults, which were growth faults during Triassic sedimentation. The Lickey End Fault continues north-westwards to join the Western Boundary Fault of the South Staffordshire Coalfield (Whitehead and Pocock, 1947); it continues southwards across the Stratford upon Avon district as the Inkberrow–Haselor Hill Fault (Williams and Whittaker, 1974, fig.3).

The Basin is an inverted early-Permian horst formed of Lower Palaeozoic or Precambrian rocks. This 'Worcester Horst' (Wills, 1956, fig.14) was a source of clasts to the Clent Breccias, which were deposited as fans in a fault-bounded basin now forming the Lickey Ridge.

The magnitude of the Triassic growth faulting is uncertain because there is no complete section through the Triassic sequence, either within the Basin or on the Lickey Ridge. The best evidence for such faulting is provided by the variation in thickness of the Bromsgrove Sandstone (Figure 6) and by the large negative gravity anomaly in the south-west of the district. In the district to the south, seismic reflection shows an increase in thickness in post-Lower Palaeozoic sediments from about 0.8 to 3 km, mainly due to growth faulting on the Weethley and Inkberrow faults (Figure 19).

Post-Jurassic folding on north–south axes occurs in the south-west (Figure 18). The north–south syncline passing through Berrow Hill is the northern continuation of the Bredon Hill Syncline (Williams and Whittaker, 1974, fig.3). Its axis trends north-north-west from just north of Berrow Hill, and continues into the adjacent district as the Droitwich Syncline (Mitchell et al., 1962).

LICKEY RIDGE

The surface expression of the Lickey Ridge is clearest in the area bounded by the Blackwell and Longbridge faults, in which Palaeozoic rocks crop out. The positive gravity

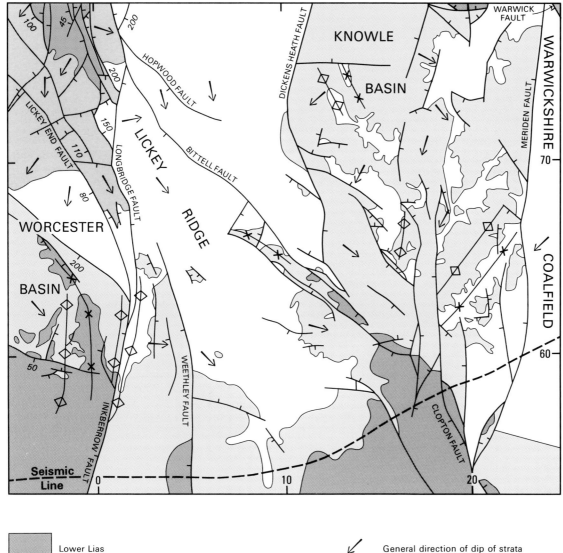

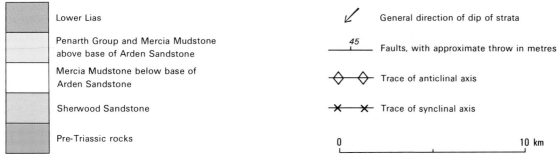

Figure 18 Principal structural elements of the district (see Figure 19 for interpretation of seismic line)

anomaly G1* coincides with the outcrop of the oldest rocks of the Lickey Inlier, and it continues northwards, corresponding with the upfaulted Palaeozoic block of the South Staffordshire Coalfield. Southwards this anomaly is expressed as a change in the anomaly gradient (G2) superimposed upon

the much more prominent negative anomaly associated with the westwards thickening of the Triassic into the Worcester Basin. The anomaly continues across the Stratford upon Avon district (Williams and Whittaker, 1974, fig.8), and connects southwards with the western margin of the gravity low at Winchcombe (Falcon and Tarrant, 1951). Within the Redditch and Stratford upon Avon districts, the trace of this anomaly approximates to that of the Lickey End–Inkberrow–Haselor Hill fault system. The anomaly is probably due to a structural basement 'high' comparable to the

---

* The gravity and magnetic anomalies referred to are shown on the Bouguer gravity and aeromagnetic anomaly inset maps on the 1:50 000 geological map.

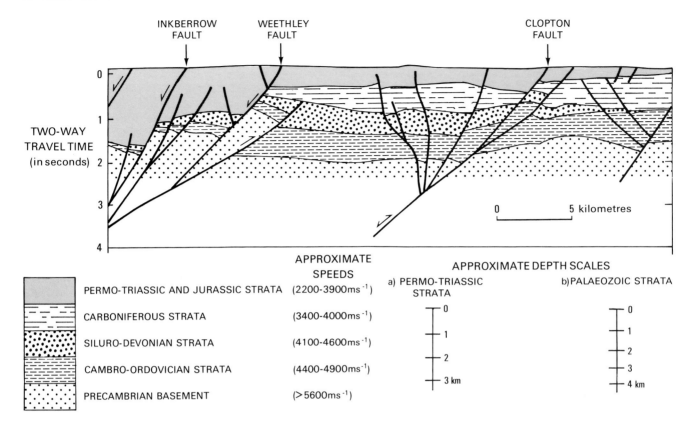

Figure 19   Geological interpretation of part of the BGS Stratford 1986 seismic line (see Figure 18 for location of seismic line)

parallel Malvern Axis 25 km to the west (Brooks, 1968), and the Vale of Moreton structure 20 km to the east (Williams and Whittaker, 1974, figs. 3 and 8).

A large positive aeromagnetic anomaly (M1–M2*) lies about 2 km east of, and parallel with the gravity anomaly just described, and is also part of an anomaly extending into the districts to the north and south. It originates in magnetised Lower Palaeozoic or Precambrian rocks which, as confirmed by the gravity anomaly, become more deeply buried southwards.

The Lickey Ridge has a complicated structural history, and there are considerable gaps in the geological sequence which originate as much from erosion following folding and faulting as from nondeposition. The older rocks of the Lickey Hills Inlier were folded before the Carboniferous, and strongly fractured by post-Triassic faulting. Dips in the Barnt Green Volcanic rocks are generally high and variable in amount, and at one locality the bedding is inverted. The beds strike north-north-west, parallel to the post-Triassic faults bounding the inlier. In the Lickey Quartzite the structures are less severe; dips are lower and more consistent within the outcrop, but since the quartzite is more competent under tectonic stress than the volcanic rocks, this does not necessarily indicate that the folding is later. The quartzite is folded overall into an anticline with its axis aligned north-north-west–south-south-east, although individual dips show that other smaller scale folds are present. At Cofton Hill [001 753] the beds are locally overfolded (Boulton, 1928).

The Lickey Quartzite is directly overlain by the Silurian at Rubery, and although the angular discordance at the base of the Rubery Sandstone is not great, the Silurian probably oversteps a considerable thickness of older Palaeozoic rocks hereabouts, indicating post-Ordovician and pre-Llandovery earth movements. The Halesowen Formation is unconformable on the Silurian and oversteps onto the Lickey Quartzite.

In contrast to the Warwickshire Coalfield, including the Coventry Horst, where the formations of the Enville Group occur in conformable sequence, the Permian Clent Breccias of the Lickey Ridge rest unconformably on the Bowhills Formation and overstep southwards onto the Keele Formation. This unconformity resulted from movements following the deposition of the Bowhills Formation, which folded the Upper Carboniferous into a gentle north-north-west-trending anticline. The breccias are coarse clastics formed as fans in a fault-bounded basin, with the 'Mercian Uplands' to the east and the 'Worcester Horst' to the west (Wills 1956, fig.14).

The gravity low G3 indicates the presence east of the Longbridge Fault of at least 600 m of sedimentary rocks overlying the Lower Palaeozoic. The relative contributions of the unproven Upper Carboniferous, Lower Permian and Sherwood Sandstone cannot be deduced, but the size of the anomaly indicates that all could be thicker than at outcrop to the west. The steepening of the gravity gradient east of anomaly G4 corresponds with the position of the Dickens Heath Fault, and may in part reflect growth faulting on the western margin of the Knowle Basin during the deposition of the Sherwood Sandstone. The anomaly G4 is positive relative to G3 and suggests some eastwards thinning of the Carboniferous against a ridge in the underlying basement. It may alternatively be due to density variations within the Lower Palaeozoic or Precambrian basement rocks, because none of the younger lithological units recognised along the Stratford seismic line (Figure 19) shows any thinning across the continuation of the anomaly, south of G5, illustrated by Williams and Whittaker (1974, fig.8). Anomaly ridge G4–G5 contrasts with the anomaly forming the axis G1–G2 in being approximately coincident with a magnetic trough (M3–M4), but the persistent north–south trend of all these anomalies indicates strong parallel structural control in the basement. The form of magnetic trough M4 suggests that there are no significant magnetic units within 10 km of the surface.

KNOWLE BASIN, INCLUDING THE WARWICKSHIRE COALFIELD

The Knowle Basin lies to the east of the Lickey Ridge and its geological history prior to Mercia Mudstone times is not well known in this district. As described above, the western margin of the Knowle Basin probably coincides with the Dickens Heath Fault. Along its eastern margin exploratory boreholes for coal have penetrated through to Tremadocian rocks (Old et al., 1987), and in the district to the north a Sherwood Sandstone sequence at least 760 m thick has been proved (Old et al., 1989). The Meriden Fault is an important structure here, but while the basin extends eastwards into the Warwick district as far as the Warwick Fault (Old et al., 1987), the Meriden Fault continues northwards as the Western Boundary Fault of the Warwickshire Coalfield (Eastwood et al., 1925).

The structural history of the eastern side of the Knowle Basin and of the Coventry Horst differs from that of the Lickey Ridge. Here, Tremadocian strata are assumed to be underlain by a full Cambrian succession similar to that proved at Nuneaton (Taylor and Rushton, 1971). The Merevale Shales of the Meer End Borehole dip at 10 to 30°, and are affected by several small faults and chevron folds. The Tremadoc is succeeded by Westphalian A rocks, any intervening strata having been overstepped. Apart from a discordance of regional magnitude at the base of the Halesowen Formation (Old et al., 1987), the Carboniferous and Permian rocks are in conformable sequence. The Lower Palaeozoic strata lie on the north-western limb of a broad north-east–south-west syncline (Old et al., 1987, fig.2), while the Carboniferous and Permian rocks lie on the western side of a broad, open, north–south-trending pre-Bromsgrove Sandstone syncline, whose western limb is truncated by the Meriden Fault (Old et al., 1987, fig.19).

The broad shallow gravity trough G6 was interpreted by Cook et al. (1951) as a Carboniferous basin extending to nearly 2 km depth, but drilling has proved that a similar anomaly centered at Leamington Spa originates in pre-Carboniferous rocks (Old et al., 1987). One possible conclu-

sion to be drawn from the similarity of the gravity anomaly pattern over the Warwickshire Coalfield of the Warwick district, and that of the eastern part of the Redditch district, is that the Productive Coal Measures continue at depth as far west as anomaly ridge G4–G5. In the district to the south, Carboniferous strata continue west to the Weethley Fault (Figure 19).

The gravity anomaly gradient G7 is associated with parallel faulting extending north-north-west from Warwick, and can in part be explained by eastwards thinning in the Triassic. The Meriden Fault, however, causes little disturbance to the gravity contours over most of its length, despite the presence of Triassic growth faulting in the Sherwood Sandstone as revealed by a seismic survey near Meer End (Allsop, 1981). Its throw must, therefore, diminish southwards, and it was not detected by the Stratford seismic line. In contrast, the throw of the *en échelon* Clopton Fault diminishes rapidly northwards.

Along magnetic ridge M5 the magnetised rocks probably do not approach within 1 km of the surface. A west-south-west–east-north-east discontinuity at depth is suggested by the alignment of the magnetic anomaly contours, and it possibly continues through the constriction in the magnetic contours between M3 and M4, and in the gravity contours between G4 and G5. An apparent east-south-east–west-north-west trend in the magnetic anomaly map connects the highs at M1 and M7, and continues east-south-east to beyond Kenilworth (Old et al., 1987), but the only surface expression of this is a sporadic weak parallelism in post-Triassic faults. The group of short wavelength magnetic highs near M7 suggests localised igneous bodies approaching closer to the surface.

There are no boreholes penetrating the full Triassic sequence west of the Meriden Fault, and the southern margin of the Knowle Basin is not clearly defined. It may lie at about grid line 70, south of which there is a slackening of the bounding gravity gradients in the west (see inset figure on 1:50 000 map), and in the east (Old et al., 1987, fig.21).

Intra- and post-Jurassic fold and fault structures in the basin are quite gentle, and the important north–south fold axes identified in the district to the south (Williams and Whittaker, 1974, fig.3) do not extend far, if at all, into this area. In particular, the Vale of Moreton Anticline, which had a marked effect on Jurassic sedimentation to the south (Whittaker, 1972), dies out in the faulted area around Bearley. Even so, syn-sedimentary movements on these faults, perhaps associated with the Moreton Anticline, produced thinning of the Blue Anchor Formation to Lower Lias sequence at Round Hill (see chapters 6 and 7).

COVENTRY HORST

The Coventry Horst is bounded in the west by the Warwick Fault which cuts across the extreme north-east corner of the district. The westwards thickening of the Triassic across this and the adjacent Meriden Fault, gives rise to gravity anomaly G7. Further description of this block is given by Old et al. (1987).

TEN

Economic geology

BRICK CLAYS

In past centuries, bricks have been manufactured locally from small workings in Mercia Mudstone, Lower Lias clays and glacial clays. Until recently, there were workings on a larger scale in the Mercia Mudstone at Knowle [166 766], Redditch [033 672] and Studley [063 637].

BUILDING STONE

The Lickey Quartzite, has been used locally for rough masonry and walling-stone. The Bromsgrove Sandstone, especially the Finstall Member, has been an important local source of stone for building purposes. Most of the workings are located just west of the present district, around Bromsgrove, but there are also sandstone workings near Tutnall [99 69] and Finstall [9763 7028]. The Bromsgrove Sandstone has been worked for railway construction at Pike's Pool [9836 7094] and in the railway cuttings at Longbridge and Cofton Hackett (Robertson and McCallum, 1930, p.48). Notable among local buildings constructed from Bromsgrove Sandstone are the parish churches at Bromsgrove and Tardebigge.

The Arden Sandstone has been widely quarried for domestic and church building (Matley, 1912, p.266), although none of the quarries was very large, perhaps because the suitable beds of sandstone are lenticular and not laterally extensive. Fine examples of the architectural use of Arden Sandstone are to be found in the churches at Rowington [2040 6927], Tanworth in Arden [1136 7052] and Beaudesert [1530 6604], and in the medieval manor house at Baddesley Clinton [1995 7146].

Limestone from the Lias Group was formerly quarried near Aston Cantlow [15 59] and was used locally for domestic architecture and, together with Arden Sandstone, in Morton Bagot church [1127 6473].

COAL

The Meriden Fault forms the western boundary to the South Warwickshire Prospect, a coalfield completely concealed beneath younger rocks. All the recoverable reserves are in the Thick Coal (Middle Coal Measures), and isopachytes and nomenclature for this seam have been published by British Coal (National Coal Board, 1985; British Coal Corporation, 1987) and by Old et al. (1987). South of Haseley [23 68], the Thick Coal lies at more than 1100 m below OD, the current limiting depth for working. The coal would provide excellent domestic, industrial and power station fuels.

The BGS Stratford 1986 seismic line indicates the presence of Carboniferous strata as far west as the Weethley Fault (Figure 19), and the possibility of Coal Measures occurring beneath the Knowle Basin is discussed in Chapter 9.

GYPSUM

Gypsum as veins and nodules is of common occurrence at various horizons in the Mercia Mudstone below the zone of weathering (Figure 9). There is a long history of gypsum working at Spernall. 'Plaster pits' existed here in the 17th century, and there were active workings via at least one adit [1018 6267] in the early part of this century. The workings are in a bed lying about 25 m below the Arden Sandstone (see p.32). The BGS Knowle Borehole [1883 7777] proved bedded gypsum/anhydrite for 3.5 m at 51 m below the base of the Penarth Group. This horizon may correlate with the Tutbury Gypsum of the Burton upon Trent district, which lies about 5 m higher stratigraphically (Stevenson and Mitchell, 1955).

LIME

Much of the Lower Lias outcrop at Knowle is covered by disused shallow pits where the limestone was formerly dug for lime, and there used to be a limeworks near Waterfield Farm [1895 7805] just to the north.

ROADSTONE

There are many small roadstone quarries in the outcrop of the Lickey Quartzite; all are now disused.

MOULDING SAND

The Wildmoor Sandstone near Bromsgrove is a major source of naturally bonded moulding sand, although the currently active works at Wildmoor [955 760] lie just beyond the present district. Former moulding sand workings at Chadwick Farm [965 755] and Bellevue [972 747] are now filled in.

SAND AND GRAVEL

The principal active sand and gravel workings are sited on the basal conglomerate of the Kidderminster Formation at Marlbrook [9824 7475] and on Kidderminster Formation sandstone at Shepley [9844 7310], the latter mainly producing building sand.

Glacial sand and gravel is currently worked on a small scale at Rowney Green [0442 7119] and Trueman's Heath

[096 765], and formerly there were workings on a larger scale at Money Lane [962 768], Lowan's Hill Farm [033 688] and Tattle Bank [177 634]. The periglacial flood gravels and river terrace deposits have not been worked on any significant scale.

The resources of sand and gravel are large and details of the Quaternary deposits, together with hectarages of possible workable mineral, are given in four special Open File Reports covering most of the district (see p.68). Two further reports (Cannell, 1982; Cannell and Crofts, 1984) deal with the assessment of these resources in the eastern half of the district. In the latter area, a range of geophysical techniques was tested during an initial feasibility study (Clarke et al., 1982) to determine practical methods of assessing small scattered drift deposits. Of these techniques ground resistivity surveying, and in particular the Offset Wenner system, was found to be the most cost-effective method of obtaining rapid and consistent results, and was successfully employed in assessing the mineral resources in the east of the district (Cannell and Crofts, 1984). A review of the mapping and assessment of aggregate resources, including those of this district, is given by Ambrose et al. (1987).

WATER SUPPLY

The district lies mostly within Hydrometric Area 54, although the northern part is within Hydrometric Area 28, and the water resources are administered by the Severn-Trent Water Authority. The principal rivers are the Alne and the Arrow which flow to the south, and there are a number of smaller streams draining to the west, north and east. The average annual rainfall varies from about 690 mm in the south-east to about 725 mm in the north (Monkhouse and Richards, 1982), and the mean annual evaporation is about 450 mm. The district has been described (Anon, 1960) as being rather dry and not lending itself to the development of large surface water supplies. Potential reservoir sites are shallow, and their construction would involve extensive flooding of farmland. A few small reservoirs are currently in use, mainly feeding canals, notably at Upper and Lower Bittell [020 750], Tardebigge [985 685] and Earlswood Lakes [11 74].

None of the Lower Palaeozoic rocks contain groundwater in significant quantities and there has been no attempt to develop the limited sources in the Upper Carboniferous.

The Triassic sandstones are the main source of groundwater, around their outcrops in the north-west and elsewhere beneath shallow Mercia Mudstone cover. In the Trias, the Kidderminster Formation, Wildmoor Sandstone and Bromsgrove Sandstone generally form a single aquifer, although it may contain aquicludes in places. The Severn-Trent Water Authority have delimited a particularly effective aquiclude just north of the district, at the top of the Kidderminster Formation, which causes groundwater head differences of more than 20 m. There are five pumping stations contributing to public supply in the west of the district, at Washingstocks, Burcot, Brockhill, Sugarbrook and Webheath (Figures 4 and 6), abstracting about 11 million cubic metres (m^3) annually; the licensed abstraction varies from about 9000 to nearly 14000 cubic metres per day (m^3d) for each source.

In the south-east there are three pumping stations, at Heath End [2324 6084], Shrewley [2214 6783] and Rowington [2092 6877], which take water from boreholes in the confined Triassic sandstones, here covered by thicknesses of 130 to 230 m of Mercia Mudstone. Approximately 1.5 million m^3 are abstracted annually from these three sites.

Early sources of water for public supply came from shafts up to 3 m in diameter, excavated in Triassic sandstones at outcrop. At Longbridge [9972 7755] and Burcot [9848 7162] the shafts were extended by headings. At a later stage, boreholes of 100 to 250 mm diameter were drilled through the bottom of the shafts, and later still, boreholes were sometimes constructed from ground level in the vicinity of the shafts, usually at a larger diameter (460 to 610 mm). The use of drilled boreholes avoids the technical difficulties and much of the expense of deeply excavated shafts, and modern sources at outcrop and in the confined aquifer are exploited in that way. Boreholes at public supply stations vary from 300 to 610 mm in diameter. The greatest depth of public supply boreholes at outcrop is about 200 m, but in the confined aquifer it is often more; at Sugarbrook [9616 6816], for example, one borehole is 398 m deep. Although most of the storage in the Triassic sandstones is intergranular, much of the groundwater flow may be through fissures, and the yield from a borehole may depend to a large extent on intersecting these.

Most of the boreholes penetrating the Triassic sandstones are for public supply; one of the few exceptions is at Barnsley Hall Hospital [9600 7258] where a borehole of 250 mm diameter and 107 m depth yielded, on test, 766 m^3/d. The mean yield of a 300 mm-diameter borehole penetrating 100 m of the saturated sandstone in the outcrop is of the order of 1000 m^3/d for a drawdown of about 25 m. Few boreholes have failed to obtain any supply. Yields from the confined aquifer near the outcrop are similar, although the drawdowns tend to be greater. In the confined aquifer in the south-east a similar borehole would yield on average about 800 m^3/d for a drawdown of 60 m or more. Lining tubes are generally required near the surface to prevent borehole collapse and the entry of surface drainage, and are also needed through the Mercia Mudstone where it overlies the sandstones. Sand screens are generally unnecessary since the sandstone usually stands without support.

In the aquifer outcrop, the groundwater quality is usually good, the total hardness (as $CaCO_3$) being generally between 100 and 300 milligrams per litre (mg/l); this is largely carbonate hardness, and the chloride ion concentration does not usually exceed 30 mg/l. The nitrate concentration can locally exceed 50 mg/l (as NO_3), but it is rare to find significant concentrations of nitrate in the confined aquifer, or even beneath a thin drift cover. The water type is usually calcium bicarbonate. In the confined aquifer, the total hardness can be more than 300 mg/l, largely due to an increase in the concentration of sulphate (SO_4). The chloride ion concentration may also rise to more than 100 mg/l, and rarely to more than 500 mg/l, particularly beneath a thick overburden of Mercia Mudstone. The groundwater tends towards the calcium sulphate and sodium chloride types.

The Mercia Mudstone is not generally considered to be an aquifer and the mudstones yield little water, but the thin sandstones or skerries can yield small supplies. Richardson (1928) noted that, in Warwickshire, excavated shafts support small demands, and that rather larger supplies can be got from the Arden Sandstone. Most of the groundwater flow in the Mercia Mudstone is through fissures in the skerries. The yield of a borehole, therefore, depends on how many skerries and fissures are penetrated, and yields vary widely. Natural replenishment of the groundwater in the skerries, particularly in the thinner ones, is often restricted; thus, although initial yields may be good, they tend to diminish with time.

An analysis of 54 boreholes drilled for water in the Mercia Mudstone suggests that on average, they are about 40 m deep and about 150 mm in diameter. The mean yield of such a borehole is about 14 m^3/d for a drawdown of 5 m. For the same drawdown, there is a 75 per cent probability of the yield exceeding 4 m^3/d, and a 25 per cent probability of exceeding 50 m^3/d. Better yields are more likely if the borehole penetrates the Arden Sandstone. In most cases, boreholes stand without support, and sand screens are rarely used. The uppermost 10 to 15 m of a borehole requires a plain lining-tube to prevent collapse of weathered ground near the surface and to avoid contamination.

Groundwater in the Mercia Mudstone tends to be hard, ranging from 200 to more than 2000 mg/l, with a mean of about 550 mg/l. When the total hardness exceeds 300 mg/l, it is mostly in the form of high concentrations of sulphate. Although saliferous beds may be present at depth beneath the western margin of the district, saline water has not generally been encountered in water boreholes (but see p.30), and the chloride ion concentration is generally less than 50 mg/l. The groundwater type varies from calcium bicarbonate to calcium sulphate.

Hydrogeologically, the most important of the superficial deposits are the glacial sands and gravels. Where these are sufficiently thick, they yield useful supplies of groundwater. A source near Haseley [237 684], comprising a well and collector trench system, yields about 1.5 million m^3 annually and more might be obtained. In the south and west, fans of periglacial flood gravel have some potential for small local supplies. River terrace gravels are generally a limited source since the water tends to drain away rapidly and long-term storage is restricted. Where sands and gravels overlie an aquifer, such as the Triassic sandstones, they are rarely used for supply since the groundwater levels are usually at a depth sufficient to leave the superficial deposits dry.

REFERENCES

Most of the references listed below are held in the Library of the British Geological Survey at Keyworth, Nottingham. Copies of the references can be purchased subject to the current copyright legislation.

ALLEN, J R L. 1965. Fining-upwards cycles in alluvial successions. *Geological Journal*, Vol. 4, 229–246.

ALLSOP, J M. 1981. Geophysical appraisal of some geological problems in the English Midlands. *Report of the Deep Geology Unit, Institute of Geological Sciences*, No. 81/7.

AMBROSE, K. 1984. *Geological notes and local details for 1:10 000 sheets SO96SE and 95NE(N), (Hanbury and Stock Green).* (Keyworth, Nottingham: British Geological Survey.)

— 1988. *Geological notes and local details for 1:10 000 sheets SO95SW and SO95SE (western part), (White Ladies Aston).* (Keyworth, Nottingham: British Geological Survey.)

— CANNELL, B, and MOORLOCK, B S P. 1987. The mapping and assessment of aggregate resources in the south Midlands and Welsh Borderland. 347–353 *in* Planning and engineering geology. CULSHAW, M G, BELL, F G, CRIPPS, J C, and O'HARA, M (editors). *Engineering Geology Special Publication of the Geological Society of London*, No. 4.

ANON. 1960. *River Severn Hydrological Survey: Hydrometric Area No. 54.* 134 pp. (London: HMSO for Ministry of Housing and Local Government.)

ARBER, E A N. 1909. On the affinities of the Triassic plant *Yuccites vogesiacus*, Schimper & Mougeot. *Geological Magazine*, Vol. 46, 11–14.

ARKELL, W J. 1933. *The Jurassic System in Great Britain.* (Oxford: Clarendon Press.)

ARTHURTON, R S. 1980. Rhythmic sedimentary sequences in the Triassic Keuper Marl (Mercia Mudstone Group) of Cheshire, northwest England. *Geological Journal*, Vol. 15, 43–58.

AUDLEY–CHARLES, M G. 1970. Triassic palaeogeography of the British Isles. *Quarterly Journal of the Geological Society of London*, Vol. 126, 49–89.

BALL, H W. 1980. *Spirorbis* from the Triassic Bromsgrove Sandstone Formation (Sherwood Sandstone Group) of Bromsgrove, Worcestershire. *Proceedings of the Geologists' Association*, Vol. 91, 149–154.

BARTON, M E. 1960. Pleistocene geology of the country around Bromsgrove. *Proceedings of the Geologists' Association*, Vol. 71, 139–155.

BASSETT, M G. 1974. Review of the stratigraphy of the Wenlock Series in the Welsh Borderland and South Wales. *Palaeontology*, Vol. 17, 745–777.

BENTON, M J. 1990. The species of *Rhynchosaurus*, a rhynchosaur (Reptilia, Diapsida) from the Middle Triassic of England. *Philosophical Transactions of the Royal Society of London*, B, Vol. 328, 213–306.

BOULTON, W S. 1925. New exposures in the Rubery-Longbridge-Rednal district, south of Birmingham: the Coal Measures and Triassic rocks. *Proceedings of the Birmingham Natural History and Philosophical Society*, Vol. 15, 61–66.

— 1928. The geology of the northern part of the Lickey Hills, near Birmingham. *Geological Magazine*, Vol. 65, 255–266.

— 1933. The rocks between the Carboniferous and Trias in the Birmingham District. *Quarterly Journal of the Geological Society of London*, Vol. 89, 53–86.

BRIDGLAND, D R, KEEN, D H, and MADDY, D. 1989. The Avon terraces. Cropthorne, Ailstone and Eckington. 51–67 *in The Pleistocene of the West Midlands: field guide.* KEEN, D H (editor). (Cambridge: Quaternary Research Association.)

BRITISH COAL CORPORATION 1987. Proposed colliery at Hawkhurst Moor. (British Coal: Central Area.)

BRITISH GEOLOGICAL SURVEY. 1984. Warwick. England and Wales Sheet 184. 1:50 000. (Southampton: Ordnance Survey for British Geological Survey.)

— 1985. Pre-Permian geology of the United Kingdom (South). 1:1 000 000. (Southampton: Ordnance Survey for British Geological Survey.)

BRODIE, P B. 1845. *A history of the fossil insects in the secondary rocks of England.* (London: John van Voorst.)

— 1856. On the Upper Keuper Sandstone (included in the New Red Marl) of Warwickshire. *Quarterly Journal of the Geological Society of London*, Vol. 12, 374–376.

— 1860. Cheirotherienfährten im oberen Keuper von Warwickshire. *Neues Jahrbuch für Mineralogie*, No. 493.

— 1865. On the Lias outliers at Knowle and Wootton Warwen in south Warwickshire, and on the presence of the Lias or Rhaetic Bone-bed at Copt Heath, its furthest northern extension hitherto recognised in that county. *Quarterly Journal of the Geological Society of London*, Vol. 21, 159–161.

— 1868. A sketch of the Lias generally in England, and of the 'Insect and Saurian Beds', especially in the lower division in the Counties of Warwick, Worcester and Gloucester, with a particular account of the fossils which characterise them. *Proceedings of the Warwickshire Naturalists' and Archaeologists' Field Club*, 1–24.

— 1874. Notes on a railway-section of the Lower Lias and Rhaetics between Stratford-on-Avon and Fenny Compton, on the occurrence of the Rhaetics near Kineton, and the insect-beds near Knowle, in Warwickshire, and on the recent discovery of the Rhaetics near Leicester. *Quarterly Journal of the Geological Society of London*, Vol. 30, 746–749.

— 1875. The Lower Lias at Ettington and Kineton and on the Rhaetics in that neighbourhood and their further extension in Leicestershire, Nottinghamshire, Lincolnshire, Yorkshire, and Cumberland. *Warwickshire Natural History and Archaeological Society Annual Report*, No. 39, 6–17

— 1893. On some additional remains of Cestraciont and other fishes in the green gritty marls, immediately overlying the red marls of the Upper Keuper in Warwickshire. *Quarterly Journal of the Geological Society of London*, Vol. 49, 171–174.

BROOKS, M. 1968. The geological results of gravity and magnetic surveys in the Malvern Hills and adjacent districts. *Geological Journal*, Vol. 6, 13–30.

BURGESS, I C, and HOLLIDAY, D W. 1974. The Permo-Triassic rocks of the Hilton Borehole. Westmorland. *Bulletin of the Geological Survey of Great Britain*, No. 46, 1–34.

BUTLER, A J. 1937. On Silurian and Cambrian rocks encountered in a deep boring at Walsall, South Staffordshire. *Geological Magazine*, Vol. 74, 241–257.

CANNELL, B. 1982. The sand and gravel resources of the country east of Solihull, Warwickshire: description of parts of 1:25 000 sheets SP 17, 18, 27 and 28. *Mineral Assessment Report of the Institute of Geological Sciences*, No.115.

— and CROFTS, R G. 1984. The sand and gravel resources of the country around Henley-in-Arden, Warwickshire: description of 1:25 000 Sheet SP 16, and parts of SP 15,17,25,26 and 27. *Mineral Assessment Report of the Institute of Geological Sciences*, No.142.

CHADWICK, R A, and SMITH, N J P. 1988. Evidence of negative structural inversion beneath central England from new seismic reflection data. *Journal of the Geological Society of London.* Vol. 145, 519–522.

CLARK, M R, BOOTH, S J, CANNELL, B, CARRUTHERS, R M, and CROFTS, R G. 1982. The resource assessment of scattered drift deposits—a feasibility study in the Redditch–Solihull area. *Mineral Assessment Unit of the Institute of Geological Sciences, Internal Report*, No. 82/1.

CLARKE, R F A. 1965. Keuper miospores from Worcestershire, England. *Palaeontology*, Vol. 8, 294–321.

CLEMENTS, R G (editor). 1977. The geology of Parkfield Road Quarry, Rugby. Unpublished report, Department of Geology, University of Leicester. 54pp.

COLLINSON, J D. 1986. Alluvial sediments. 20–62 in *Sedimentary environments and facies* (2nd edition). READING, H G (editor). (Oxford: Blackwell Scientific Publications.)

COOK, A H, HOSPERS, J, and PARASNIS, D S. 1951. The results of a gravity survey in the country between the Clee Hills and Nuneaton. *Quarterly Journal of the Geological Society of London*, Vol. 107, 287–306.

COPE, J C W, GETTY, T A, HOWARTH, M K, MORTON, N, and TORRENS, H S. 1980. A correlation of Jurassic rocks in the British Isles. Part 1: Introduction and Lower Jurassic. *Special Report of the Geological Society of London*, No. 14.

COPE, K G, and JONES, A R L. 1970. The Warwickshire Thick Coal and its mining development. *Compte Rendu du Congrès International de Stratigraphie et de Géologie du Carbonifère, Sheffield*, 1967, Vol. 2, 585–598.

COUPER, R A. 1958. British Mesozoic microspores and pollen grains. *Palaeontographica B*, Vol. 103, 75–179.

COWIE, J W, RUSHTON, A W A, and STUBBLEFIELD, C J. 1972. A correlation of Cambrian rocks in the British Isles. *Special Report of the Geological Society of London*, No. 2.

DEAN, W T, DONOVAN, D T, and HOWARTH, M K. 1961. The Liassic ammonite zones and subzones of the North-West European Province. *Bulletin of the British Museum, Natural History (Geology)*, Vol. 4, 435–505.

DEMATHIEU, G, and HAUBOLD, H. 1972. Stratigraphische Aussagen der Tetrapodenfährten aus der terrestrischen Trias Europas. *Geologie*, Jg.21, H.7, 802–836.

DONOVAN, D T. 1956. The zonal stratigraphy of the Blue Lias around Keynsham, Somerset. *Proceedings of the Geologists' Association*, Vol. 66, 182–212.

— HORTON, A, and IVIMEY-COOK, H C. 1979. The transgression of the Lower Lias over the northern flank of the London Platform. *Journal of the Geological Society of London*, Vol. 136, 165–173.

DUNNING, F W. 1975. Precambrian craton of central England and the Welsh Borders. 83–95 *in* A correlation of Precambrian rocks in the British Isles. HARRIS, A L and five others (editors). *Special Report of the Geological Society of London*, No. 6.

EARP, J R, and TAYLOR, B J. 1986. Geology of the country around Chester and Winsford. *Memoir of the British Geological Survey*, Sheet 109 (England and Wales).

EASTWOOD, T, GIBSON, W, CANTRILL, T C and WHITEHEAD, T H. 1923. The geology of the country around Coventry. *Memoir of the Geological Survey of Great Britain*, Sheet 169.

— WHITEHEAD, T H, and ROBERTSON, T. 1925. The geology of the country around Birmingham. *Memoir of the Geological Survey of Great Britain*, Sheet 168.

EDMONDS, E A, POOLE, E G, and WILSON, V. 1965. Geology of the country around Banbury and Edge Hill. *Memoir of the Geological Survey of Great Britain*, Sheet 201 (England and Wales).

EGERTON, P DE M G. 1854. On a fossil fish from the upper beds of the New Red Sandstone at Bromsgrove. *Quarterly Journal of the Geological Society of London*, Vol. 10, 367–371.

— 1858. On *Palaeoniscus superstes*. *Quarterly Journal of the Geological Society of London*, Vol. 14, 164–167.

ELLIOTT, R E. 1961. The stratigraphy of the Keuper Series in southern Nottinghamshire. *Proceedings of the Yorkshire Geological Society*, Vol. 33, 197–234.

FALCON, N L, and TARRANT, L H. 1951. The gravitational and magnetic exploration of parts of the Mesozoic-covered areas of south-central England. *Quarterly Journal of the Geological Society of London*, Vol. 106, 141–170.

FITCH, F J, MILLER, J A, and THOMPSON, D B. 1966. The palaeogeographic significance of isotope age determinations on detrital micas from the Triassic of the Stockport–Macclesfield district, Cheshire, England. *Palaeogeography, Palaeoclimatology, Palaeoecology*, Vol. 2, 281–312.

FLEET, W F. 1923. Notes on the Triassic sands near Birmingham with special reference to their heavy detrital minerals. *Proceedings of the Geologists' Association*, Vol. 34, 114–119.

— 1925. The chief heavy detrital minerals in the rocks of the English Midlands. *Geological Magazine*, Vol. 62, 98–128.

GALL, J C, GRAUVOGEL, L, and LEHMAN, J P. 1974. Faune du Buntsandstein. V. Les poissons fossiles de la collection Grauvogel-Gall. *Annales de Paléontologie*, Vol. 60, 129–147.

GALTON, P M. 1985. The poposaurid thecodontian *Teratosaurus suevicus* v. Meyer, plus referred specimens mostly based on prosauropod dinosaurs, from the Middle Stubensandstein (Upper Triassic) of Nordwürttemberg. *Stuttgärter Beiträge zur Naturkunde*, Serie B, No.116, 1–29.

GEORGE, T N. 1937. The geology of the district around Dunhampstead and Himbleton, Worcestershire. *Summary of Progress of the Geological Survey of Great Britain for 1935*, pt 2, 119–129.

GIBBARD, P L, and PEGLAR, S M. 1989. Palynology of the fossiliferous deposits at Brandon, Warwickshire. 23–26 in *The Pleistocene of the West Midlands: field guide*. KEEN, D H (editor). (Cambridge: Quaternary Research Association.)

GIBSON, W, and WATTS, W W. 1898. Lickey Hills. 67–69 in *Summary of progress for 1897, Geological Survey*. (London: Her Majesty's Stationery Office.)

GLENNIE, K W, and EVANS G. 1976. A reconnaissance of the Recent sediments of the Ranns of Kutch, India. *Sedimentology*, Vol. 23, 625–647.

GRAUVOGEL-STAMM, L. 1969. Nouveux types d'organes reproducteurs mâles de conifères du Grès à *Voltzia* (Trias Inférieur) des Vosges. *Bulletin Service Carte Géologique Alsace Lorraine.* Vol. 22, 93 – 120.

— 1972. Révision de cônes mâles du 'Keuper Inférieur' du Worcestershire (Angleterre) attribués à *Masculostrobus willsi* Townrow. *Palaeontographica*, B, Vol.140, 1 – 26.

— and SCHAARSCHMIDT, F. 1978. Zur nomenklatur von *Masculostrobus* Seward. *Sciences Géologiques Bulletin*, Vol.31, 105 – 107.

— 1979. Zur Morphologie und Taxonomie von *Masculostrobus* SEWARD und anderen Formgattungen peltater männlicher Koniferenblüten. *Senckenbergiana Lethaea*, Vol. 60, 1 – 37.

GREIG, D C, WRIGHT, J E, HAINS, B A, and MITCHELL, G H. 1968. Geology of the country around Church Stretton, Craven Arms, Wenlock Edge and Brown Clee. *Memoir of the Geological Survey of Great Britain,* Sheet 166.

HALLAM, A. 1964. Origin of the limestone-shale rhythm in the Blue Lias of England. A composite theory. *Journal of Geology,* Vol. 72, 157 – 169.

— 1967. An environmental study of the Upper Domerian and Lower Toarcian in Great Britain. *Philosophical Transactions of the Royal Society of London*, B, Vol. 252, 393 – 445.

HAMBLIN, R J O, WARWICK, G T, and WHITE, D E. 1978. Geological Handbook for the Wrens Nest National Nature Reserve. *Nature Conservancy Council.* 16pp.

HARDIE, W G. 1954. The Silurian rocks of Kendal End, near Barnt Green, Worcestershire. *Proceedings of the Geologists' Association,* Vol. 65, 11 – 17.

HARRISON, R K, OLD, R A, STYLES, M T, and YOUNG, B R. 1983. Coffinite in nodules from the Mercia Mudstone Group (Triassic) of the IGS Knowle Borehole, West Midlands. *Report of the Institute of Geological Sciences*, No. 83/10, 12 – 16.

HASSAN, S M. 1964. A comparative study of the sedimentology of the Upper Mottled Sandstone and the Lower Keuper Sandstone of an area west of Birmingham. Unpublished MSc thesis, University of Birmingham.

HOLLOWAY, S, MILODOWSKI, A E, STRONG, G E, and WARRINGTON, G. 1989. The Sherwood Sandstone Group (Triassic) of the Wessex Basin, southern England. *Proceedings of the Geologists' Association*, Vol. 100, 383 – 394.

HORTON, A. 1974. The sequence of Pleistocene deposits proved during the construction of the Birmingham motorways. *Report of the Institute of Geological Sciences*, No. 74/11.

HULL, E. 1869. The Triassic and Permian rocks of the Midland counties of England. *Memoir of the Geological Survey of Great Britain.*

INSTITUTE OF GEOLOGICAL SCIENCES. 1977. Quaternary map of the United Kingdom, South. 1:625 000.

— 1982. IGS boreholes 1980. *Report of the Institute of Geological Sciences*, No. 81/11.

JACKSON, A A. 1982. *Geological notes and local details for 1:10 000 sheet: SP06SW and SE (Astwood Bank and Studley).* (Keyworth, Nottingham: Institute of Geological Sciences.)

JEANS, C V. 1978. The origin of the Triassic clay assemblages of Europe with special reference to the Keuper Marl and Rhaetic of parts of England. *Philosophical Transactions of the Royal Society of London*, A, Vol. 289, 549 – 639.

KELLY, M R. 1964. The Middle Pleistocene of north Birmingham. *Philosophical Transactions of the Royal Society of London*, B, Vol. 247, 533 – 592.

— 1968. Floras of Middle and Upper Pleistocene age, from Brandon, Warwickshire. *Philosophical Transactions of the Royal Society of London*, B, Vol. 254, 401 – 415.

KENT, P E. 1975. The tectonic development of Great Britain and the surrounding seas. 3 – 28 in *Petroleum and the Continental Shelf of North West Europe, Vol. 1, Geology.* WOODLAND, A W (editor). (London: Applied Science Publishers.)

KING, W W. 1899. The Permian conglomerates of the Lower Severn basin. *Quarterly Journal of the Geological Society of London,* Vol. 55, 97 – 128.

KJELLESVIG-WAERING, E N. 1986. A restudy of the fossil Scorpionida of the world. *Palaeontographica Americana*, No. 55, 1 – 287.

KOBAYASHI, T. 1954. Fossil Estherians and allied fossils. *Journal of the Faculty of Science, Tokyo University*, Section II, Vol. 9, 1 – 192.

LAMONT, A A. 1946. Fossils from Middle Bunter pebbles collected in Birmingham. *Geological Magazine*, Vol. 83, 39 – 44.

LAPWORTH, C. 1899. Sketch of the geology of the Birmingham district, with special reference to the long excursion of 1898. *Proceedings of the Geologists' Association*, Vol. 15, 313 – 415.

LISTER, A M, and KEEN, D H. 1989. Snitterfield Pit and faunal remains. 34 – 35 in *The Pleistocene of the West Midlands: field guide.* KEEN, D H (editor). (Cambridge: Quaternary Research Association.)

LUCY, W C. 1872. The gravels of the Severn, Avon and Evenlode and their extension over the Cotteswold Hills. *Proceedings of the Cotteswold Naturalists' Field Club*, Vol. 5, 71 – 142.

McCALLUM, R J. 1927. The opening out of Cofton Tunnel London, Midland and Scottish Railway. *Minutes and Proceedings of the Institute of Civil Engineers*, Vol. 231, paper 4773, 161 – 187.

McCUNE, A R. 1986. A revision of *Semionotus* (Pisces:Semionotidae) from the Triassic and Jurassic of Europe. *Palaeontology*, Vol. 29, 213 – 233.

MATLEY, C A. 1912. The Upper Keuper (or Arden) Sandstone Group and associated rocks of Warwickshire. *Quarterly Journal of the Geological Society of London*, Vol. 68, 252 – 280.

MIALL, A D. 1977. A review of the braided-river depositional environment. *Earth Sciences Review*, Vol. 13, 1 – 62.

MITCHELL, G F, PENNY, L F, SHOTTON, F W, and WEST, R G. 1973. A correlation of Quaternary deposits in the British Isles. *Special Report of the Geological Society of London*, No. 4.

MITCHELL, G H. 1942. The geology of the Warwickshire Coalfield. *Wartime Pamphlet of the Geological Survey of Great Britain*, No. 25.

— POCOCK, R W, and TAYLOR, J H. 1962. Geology of the country around Droitwich, Abberley and Kidderminster. *Memoir of the Geological Survey of Great Britain*, Sheet 182.

MONKHOUSE, R A, and RICHARDS, H J. 1982. Groundwater resources of the United Kingdom. Report for the Directorate-General for the Environment Consumer Protection and Nuclear Safety, Commission of the European Communities. (Hannover: Th. Schafer.)

MURCHISON, R I. 1839. *Silurian System, Part 1.* (London: John Murray.)

— and STRICKLAND H E. 1840. On the upper formations of the New Red Sandstone System in Gloucestershire, Worcestershire and Warwickshire; etc. *Transactions of the Geological Society of London*, Ser. 2, Vol. 5, 331 – 348.

NATIONAL COAL BOARD. 1985. The South Warwickshire Prospect: A consultation paper. (NCB: South Midlands Area.)

NEWTON, E T. 1887. On the remains of fishes from the Keuper of Warwick and Nottingham. *Quarterly Journal of the Geological Society of London,* Vol. 43, 537–543.

NEWTON, R B. 1893. Note on some molluscan remains lately discovered in the English Keuper. *Geological Magazine,* Vol. 40, 557.

— 1894. Note on some molluscan remains lately discovered in the English Keuper. *Journal of Conchology,* Vol. 7, 408–13.

OLD, R A. 1982. *Geological notes and local details for 1:10 000 sheet SP17NE (Solihull and Knowle).* (Keyworth, Nottingham: Institute of Geological Sciences.)

— 1983. *Geological notes and local details for 1:10 000 sheet SPO7NE (Hollywood).* (Keyworth, Nottingham: Institute of Geological Sciences.)

— SUMBLER, M G, and AMBROSE, K. 1987. Geology of the country around Warwick. *Memoir of the British Geological Survey,* Sheet 184 (England and Wales).

— BRIDGE, D McC, and REES, J G. 1989. The geology of the Coventry area. *British Geological Survey Technical Report,* WA/89/29.

OSBORNE, P J, and SHOTTON, F W. 1968. The fauna of the channel deposit of early Saalian age, at Brandon, Warwickshire. *Philosophical Transactions of the Royal Society of London,* B. Vol. 254, 417–424.

OWEN, T R. 1976. *The geological evolution of the British Isles.* (Oxford: Pergamon Press.)

PATON, R L. 1974. Capitosauroid labyrinthodonts from the Trias of England. *Palaeontology,* Vol. 17, 253–289.

PATTISON, J, SMITH, D B, and WARRINGTON, G. 1973. A review of late Permian and early Triassic biostratigraphy in the British Isles. 220–260 in The Permian and Triassic systems and their mutual boundary. LOGAN, A and HILLS, L V (editors). *Memoir of the Canadian Society of Petroleum Geologists,* No. 2.

PETTIJOHN, F J, POTTER, P E and SIEMER, R. 1972. *Sand and sandstone.* (Berlin, Heidelberg and New York: Springer Verlag.)

PHILLIPS, J. 1871. *Geology of Oxford and the valley of the Thames.* (Oxford: Clarendon Press.)

PICKERING, R. 1957. The Pleistocene deposits of the south Birmingham area. *Quarterly Journal of the Geological Society of London,* Vol. 113, 223–239.

POOLE, E G. 1969. The Putcheons Farm (1965) Borehole, Redditch, Worcestershire. *Bulletin of the Geological Survey of Great Britain,* No. 29, 105–114.

— and WILLIAMS, B J. 1981. The Keuper Saliferous Beds of the Droitwich area. *Report of the Institute of Geological Sciences,* No. 81/2.

POWELL, J H. 1984. Lithostratigraphical nomenclature of the Lias Group in the Yorkshire Basin. *Proceedings of the Yorkshire Geological Society,* Vol. 45, 51–57.

RAMSBOTTOM, W H C, CALVER, M A, EAGAR, R M C, HODSON, F, HOLLIDAY, D W, STUBBLEFIELD, C J, and WILSON, R B. 1978. A correlation of Silesian Rocks in the British Isles. *Special Report of the Geological Society of London,* No.10.

RICE, R J. 1968. The Quaternary deposits of central Leicestershire. *Philosophical Transactions of the Royal Society of London,* B. Vol. 262, 459–509.

RICHARDSON, L. 1904. The Rhaetic rocks of Worcestershire. *Proceedings of the Cotteswold Naturalists' Field Club,* Vol. 15, 19–43.

— 1905. The Lias of Worcestershire. *Transactions of the Worcester Naturalists' Club,* Vol. 3, 188–206.

— 1928. Wells and springs of Warwickshire. *Memoir of the Geological Survey of Great Britain.*

— 1930. Wells and springs of Worcestershire. *Memoir of the Geolo Great Britain,* Pt. 2, (London: Her Majesty's Stationery Office.)

ROSE, G N, and KENT, P E. 1955. A *Lingula*-bed in the Keuper of Nottinghamshire. *Geological Magazine,* Vol. 92, 476–480.

— 1989. Tracing the Baginton–Lillington Gravels from the West Midlands to East Anglia. 102–119 in *The Pleistocene of the West Midlands: field guide.* KEEN, D H (editor). (Cambridge: Quaternary Research Association.)

ROSE, J. 1987. Status of the Wolstonian glaciation in the British Quaternary. *Quaternary Newsletter,* No. 53, 1–9.

SARJEANT, W A S. 1974. A history and bibliography of the study of fossil vertebrate footprints in the British Isles. *Palaeogeography, Palaeoclimatology, Palaeoecology.* Vol. 16, 265–378.

SELLEY, R C. 1978. *Ancient sedimentary environments.* (London: Chapman and Hall.)

SHEARMAN, D J, MOSSOP, G, DUNSMORE, H, and MARTIN, W. 1972. Origin of gypsum veins by hydraulic fracture. *Transactions of the Institute of Mining and Metallurgy, B, Applied Earth Sciences,* Vol.81, 149–155.

SHOTTON, F W. 1953. The Pleistocene deposits of the area between Coventry, Rugby and Leamington, and their bearing upon the topographic development of the Midlands. *Philosophical Transactions of the Royal Society of London,* B, Vol. 237, 209–260.

— 1968a. The Pleistocene sequence in a pipe trench between Chadwick End and Gibbet Hill, South of Coventry, Warwickshire. *Proceedings of the Coventry and District Natural History and Scientific Society,* Vol. 4, 53–59.

— 1968b. The Pleistocene succession around Brandon, Warwickshire. *Philosophical Transactions of the Royal Society of London,* B, Vol. 254, 387–400.

ROBERTSON, T, and McCALLUM, R T. 1930. L.M.&S.R. Longbridge and Barnt Green widening near Birmingham. 42–53 in *Summary of progress for 1929.* GEOLOGICAL SURVEY OF GREAT BRITAIN, Pt. 2, (London: Her Majesty's Stationery Office.)

ROSE, G N, and KENT, P E. 1955. A *Lingula*-bed in the Keuper of Nottinghamshire. *Geological Magazine,* Vol. 92, 476–480.

ROSE, J. 1987. Status of the Wolstonian glaciation in the British Quaternary. *Quaternary Newsletter,* No. 53, 1–9.

— 1989. Tracing the Baginton–Lillington Gravels from the West Midlands to East Anglia. 102–119 in *The Pleistocene of the West Midlands: field guide.* KEEN, D H (editor). (Cambridge: Quaternary Research Association.)

SARJEANT, W A S. 1974. A history and bibliography of the study of fossil vertebrate footprints in the British Isles. *Palaeogeography, Palaeoclimatology, Palaeoecology.* Vol. 16, 265–378.

SELLEY, R C. 1978. *Ancient sedimentary environments.* (London: Chapman and Hall.)

SHEARMAN, D J, MOSSOP, G, DUNSMORE, H, and MARTIN, W. 1972. Origin of gypsum veins by hydraulic fracture. *Transactions of the Institute of Mining and Metallurgy, B, Applied Earth Sciences,* Vol.81, 149–155.

SHOTTON, F W. 1953. The Pleistocene deposits of the area between Coventry, Rugby and Leamington, and their bearing upon the topographic development of the Midlands. *Philosophical Transactions of the Royal Society of London,* B, Vol. 237, 209–260.

— 1968a. The Pleistocene sequence in a pipe trench between Chadwick End and Gibbet Hill, South of Coventry, Warwickshire. *Proceedings of the Coventry and district Natural History and Scientific Society,* Vol. 4, 53 – 59.

— 1968b. The Pleistocene succession around Brandon, Warwickshire. *Philosophical Transactions of the Royal Society of London,* B, Vol. 254, 387 – 400.

— 1976. Amplification of the Wolstonian stage of the British Pleistocene. *Geological Magazine,* Vol. 113, 241 – 50.

— 1985. Interpretation of an old section in Pleistocene deposits south of Temple Balsall, Warwickshire. *Proceedings of the Coventry and district Natural History and Scientific Society,* Vol. 5, 307 – 313.

— 1989. The Wolston sequence and its position within the Pleistocene. 1 – 4 in *The Pleistocene of the West Midlands: field guide.* KEEN D H (editor). (Canbridge: Quaternary Research Association.)

— and WEST, R G. 1969 Stratigraphic table of the British Quaternary. 155 – 157 in Recommendations on stratigraphical usage. *Proceedings of the Geological Society of London,* No. 1656.

SMITH, D B, BRUNSTROM, R G W, MANNING, P I, SIMPSON, S, and SHOTTON, F W. 1974. A correlation of Permian rocks in the British Isles. *Special Report of the Geological Society of London,* No. 5.

SMITH, W C. 1963. Description of the igneous rocks represented among pebbles from the Bunter Pebble Beds of the Midlands of England. *Bulletin of the British Museum (Natural History) Mineralogy,* Vol. 2, 1 – 17.

STEVENSON, I P, and MITCHELL, G H. 1955. Geology of the country between Burton upon Trent, Rugeley and Uttoxeter. *Memoir of the Geological Survey of Great Britain,* Sheet 140 (England and Wales).

STRANGE, P J, and AMBROSE, K. 1982. *Geological notes and local details for 1:10 000 sheets SP16 and parts of SP15 (Henley in Arden).* (Keyworth, Nottingham: Institute of Geological Sciences.)

STRICKLAND, A E. 1842. Memoir descriptive of a series of coloured sections of the cuttings on the Birmingham and Gloucester Railway. *Transactions of the Geological Society of London,* Series 2, Vol. 6, 545 – 555.

STRONG, G E. 1983. Petrographical notes on specimens of Lickey Hill quartzites. *Report of the Petrology Unit, Institute of Geological Sciences,* 246.

SUMBLER, M G. 1983. A new look at the type Wolstonian glacial deposits of Central England. *Proceedings of the Geologists' Association,* Vol. 94, 23 – 31.

— 1984. *Geological notes and local details for 1:10 000 sheets SP 47 NW, NE, SW, SE (Rugby west).* (Keyworth, Nottingham: British Geological Survey.)

TAYLOR, K, and RUSHTON, A W A. 1971. The pre-Westphalian geology of the Warwickshire Coalfield. *Bulletin of the Geological Survey of Great Britain,* No. 35.

TOMES, R F. 1878. On the stratigraphical position of the corals of the Lias of the Midland and Western Counties of England and South Wales. *Quarterly Journal of the Geological Society of London,* Vol. 34, 179 – 195.

TOMLINSON, M E. 1925. River terraces of the lower valley of the Warwickshire Avon. *Quarterly Journal of the Geological Society of London,* Vol. 81, 137 – 169.

— 1929. The drifts of the Stour-Evenlode watershed and their extension into the valleys of the Warwickshire Stour and upper Evenlode. *Proceedings of the Birmingham Natural History and Philosophical Society,* Vol. 15, 157 – 195.

— 1935. The superficial deposits of the country north of Stratford on Avon. *Quarterly Journal of the Geological Society of London,* Vol. 91, 423 – 462.

TOWNROW, J A. 1962. On some disaccate pollen grains of Permian to middle Jurassic age. *Grana Palynologica,* Vol. 3, 13 – 44.

TRUSHEIM, F. 1963. Zur Gliederung des Buntsandsteins. *Erdöl-Zeitschrift für Bohr- und Fördertechnik,* Vol. 79, 277 – 292.

WALKER, A D. 1969. The reptile fauna of the "Lower Keuper" Sandstone. *Geological Magazine,* Vol. 106, 470 – 476.

WALKER, T R. 1976. Diagenetic origin of continental red beds. 240 – 282 in *The continental Permian in Central, West and Southern Europe,* FALKE, N (editor). (Dordrecht: Holland.)

WARRINGTON, G. 1967. Correlation of the Keuper Series of the Triassic by Miospores. *Nature, London,* Vol. 214, 1323 – 1324.

— 1968. The stratigraphy and palaeontology of the "Keuper" Series (upper Triassic) in central England (Worcestershire, Warwickshire, Stafford and Leicestershire). Unpublished PhD thesis, University of London.

— 1970. The stratigraphy and palaeontology of the "Keuper" Series of the central Midlands of England. *Quarterly Journal of the Geological Society of London,* Vol. 126, 183 – 223.

— 1974. Les évaporites du Trias britannique. *Bulletin de la Société Géologique de la France,* Series 7, Vol. 16, No. 6, 708 – 723.

— 1981. The indigenous micropalaeontology of British Triassic shelf sea deposits. 61 – 70 in *Microfossils from recent and fossil shelf seas.* NEALE, J W, and BRASIER, M D (editors). (Horwood: Chichester.)

— AUDLEY-CHARLES, M G, ELLIOTT, R E, EVANS, W B, IVIMEY-COOK, H C, KENT, P E, ROBINSON P L, SHOTTON, F W, and TAYLOR, F M. 1980. A correlation of Triassic rocks in the British Isles. *Special Report of the Geological Society of London,* No. 13.

— and IVIMEY-COOK, H C. In press. Triassic. In *Atlas of palaeogeography and lithofacies.* (Geological Society of London.)

WELIN, E, ENGSTRAND, L, and VACZY, S. 1975. Institute of Geological Sciences radiocarbon dates VI. *Radiocarbon,* Vol. 17, 158.

WHITE, E. 1950. A fish from the Bunter near Kidderminster. *Transactions of the Worcestershire Naturalists' Club,* Vol. 10, 185 – 189.

WHITEHEAD, T H, and POCOCK, R W. 1947. Dudley and Bridgnorth. *Memoir of the Geological Survey of Great Britain,* Sheet 167 (England and Wales).

WHITTAKER, A. 1972. Intra-Liassic structures in the Severn Basin area. *Report of the Institute of Geological Sciences,* No. 72/3.

— (editor). 1985. *Atlas of onshore sedimentary basins in England and Wales: post-Carboniferous tectonics and stratigraphy.* (Glasgow and London: Blackie.)

WILLIAMS, B J, and WHITTAKER, A. 1974. Geology of the country around Stratford-upon-Avon and Evesham. *Memoir of the Geological Survey of Great Britain,* Sheet 200 (England and Wales).

WILLE, W. 1970. *Plaesiodictyon mosellanum* n.g.,n.sp., eine mehrzellige Grünalge aus dem Unteren Keuper von Luxemburg. *Neues Jahrbuch für Geologie und Paläontologie Monatshefte,* 5, 283 – 310.

WILLS, L J. 1907a. On some fossiliferous Keuper rocks at Bromsgrove (Worcestershire). *Geological Magazine,* Vol. 44, 28 – 34.

— 1907b. Note on the fossils from the Lower Keuper of Bromsgrove. *Report of the British Association,* 312 – 313.

— 1910a.   On the fossiliferous lowerKeuper rocks of Worcestershire, with descriptions of some of the plants and animals discovered therein.   *Proceedings of the Geologists' Association,* Vol. 21, 249–331.

— 1910b.   Notes on the genus *Schizoneura,* Schimper and Mougeot.   *Proceedings of the Cambridge Philosophical Society,* Vol. 15, 406–410.

— 1916.   The structure of the lower jaw of Triassic labyrinthodonts.   *Proceedings of the Birmingham Natural History and Philosophical Society.* Vol. 14, 5–20.

— 1937.   The Pleistocene history of the West Midlands.   *Report of the British Association for the Advancement of Science, Nottingham,* C, 71–94.

— 1938.   The Pleistocene development of the Severn from Bridgnorth to the sea.   *Quarterly Journal of the Geological Society of London,* Vol. 94, 161–242.

— 1945.   Field meeting in the Lickey Hills.   *Proceedings of the Geologists' Association,* Vol. 56, 23–25.

— 1947.   *A monograph of British Triassic scorpions.*   Part I, 1–74, Part II, 75–137. (London: The Palaeontographical Society.)

— 1948.   *The palaeogeography of the Midlands.*   (University Press of Liverpool.)

— 1956.   *Concealed coalfields.*   (London: Blackie.)

— 1970a.   The Triassic succession in the central Midlands in its regional setting.   *Quarterly Journal of the Geological Society of London,* Vol. 126, 225–285.

— 1970b.   The Bunter Formation at the Bellington Pumping Station of the East Worcestershire Waterworks Company. *Mercian Geologist,* Vol. 3, 387–397.

— 1976.   The Trias of Worcestershire and Warwickshire. *Report of the Institute of Geological Sciences,* No. 76/2.

— and CAMPBELL–SMITH W. 1913.   Notes on the flora and fauna of the Upper Keuper Sandstones of Warwickshire and Worcestershire.   *Geological Magazine,* Vol. 50, 461–462.

— and LAURIE, W H. 1938.   Deep sewer trench along the Bristol Road from Ashill Road near the Longbridge Hotel to the city boundary at Rubery, 1937.   *Proceedings of the Birmingham Natural History and Philosophical Society,* Vol. 16, 175–180.

— and SHOTTON, F W. 1938.   A quartzite breccia at the base of the Trias exposed in a trench in Tessall Lane, Northfield.

Proceedings of the Birmingham Natural History and Philosophical Society, Vol. 16, 181–183.

— and SARJEANT, W A S. 1970.   Fossil vertebrate and invertebrate tracks from boreholes through the Bunter Series (Triassic) of Worcestershire.   *Mercian Geologist,* Vol. 3, 399–414.

— WILKINS, L G, and HUBBARD, G H. 1925.   The Upper Llandovery Series of Rubery.   *Proceedings of the Birmingham Natural History and Philosophical Society,* Vol. 15, 67–83.

WOODWARD, H B. 1887.   *The geology of England and Wales with notes on the physical features of the country.*   (London: G Philip.)

— 1893.   The Jurassic rocks of Britain: Vol. III. The Lias of England and Wales (Yorkshire excepted).   *Memoir of the Geological Survey of Great Britain.*

WORSSAM, B C, and OLD, R A. 1988.   Geology of the country around Coalville.   *Memoir of the British Geological Survey,* Sheet 155 (England and Wales).

— ELLISON, R A, and MOORLOCK, B S P. 1989.   Geology of the country around Tewkesbury.   *Memoir of the British Geological Survey,* Sheet 216 (England and Wales).

WRIGHT, T. 1860.   On the zone of *Avicula contorta* and the Lower Lias of the south of England.   *Quarterly Journal of the Geological Society of London,* Vol. 16, 374–411

— 1878–1886.   Monograph of the Lias ammonites of the British Islands.   *Monograph of the Palaeontographical Society,* 1–503.

WURSTER, P. 1964.   Delta sedimentation in the German Keuper Basin.   436–446 in *Developments in sedimentology, Vol. 1. Deltaic and shallow marine deposits.* VAN STRAATEN, L M J U (editor). (London, Amsterdam and New York: Elsevier.)

WYATT, R J, and MOORLOCK, B S P. 1982.   *Geological notes and local details for 1:10 000 sheets SO83NE (Ripple).*   (Keyworth, Nottingham: Institute of Geological Sciences.)

YATES, J. 1829.   Observations on the structure of the border country of Salop and North Wales; and of some detached groups of Transition Rocks in the Midland Counties.   *Transactions of the Geological Society of London,* Series 2, Vol. 2, 237–264.

ZIEGLER, A M, COCKS, L R M, and MCKERROW, W S. 1968. The Llandovery transgression of the Welsh Borderland. *Palaeontology.* Vol. 11, 736–782.

APPENDIX 1

1:10 000 maps

Geological National Grid 1:10 000 maps included wholly or in part in Sheet 183 are listed below, together with the initials of the surveyors and the date of survey.

The surveyors were K Ambrose, R J O Hamblin, A Horton, A A Jackson, R A Old and B C Worssam for Sheet 183 and E G Poole and B J Williams for the peripheral areas.

SO 95 NE	Kington	BJW	1966
		KA	1982
SO 96 NE	Stoke Prior	RJOH	1980 – 81
SO 96 SE	Hanbury	KA, RJOH	1982
SO 97 NE	Rubery	RJOH	1980
SO 97 SE	Bromsgrove	RJOH	1979 – 80
SP 05 NW	Inkberrow	BJW	1966
		AAJ	1980
SP 05 NE	Alcester	BJW	1966
		AAJ	1980
SP 06 NW	Redditch West	AH	1970 – 71
SP 06 NE	Redditch East	AH, BCW	1969 – 70
SP 06 SW	Astwood Bank	AH	1971 – 72
		AA	1979 – 80
SP 06 SE	Studley	AH	1971
		BCW	1972
		AAJ	1980
SP 07 NW	Longbridge	RJOH	1979, 1981
SP 07 NE	Hollywood	RAO	1982
SP 07 SW	Alvechurch	RJOH	1979 – 80
SP 07 SE	Inkford	KA	1982
SP 15 NW	Great Alne	BJW	1965
		PJS	1980
		KA	1984
SP 15 NE	Wilmcote	BJW	1964
		PJS	1980
		KA	1984
SP 16 NW	Ullenhall	PJS	1980
SP 16 NE	Henley-in-Arden	KA	1980
SP 16 SW	Morton Bagot	PJS	1980
		KA	1984
SP 16 SE	Bearley	PJS	1979
		KA	1984
SP 17 NW	Shirley	RAO	1981
SP 17 NE	Solihull and Knowle	RAO	1980
SP 17 SW	Tanworth-in-Arden	KA	1982
SP 17 SE	Lapworth	KA	1981
SP 25 NW	Tiddington	EGP	1957
		BJW	1964
		KA	1978
		PJS	1979
SP 26 NW	Shrewley	KA	1977
SP 26 SW	Norton Lindsey	PJS	1979
SP 27 NW	Berkswell and Balsall Common	RAO	1978 – 80
SP 27 SW	Wroxall	RAO	1978 – 79

APPENDIX 2

Open file reports

The open file reports listed below are detailed accounts of the geology of the constituent 1:10 000 sheets of the Redditch (183) 1:50 000 sheet. The full title is given for the first report; the others all carry similar titles. Copies of the reports may be ordered from the British Geological Survey, Keyworth, Nottingham.

SO 96 SE (and part of 95 NE) Ambrose, K. 1984. *Geological notes and local details for 1:10 000 sheets SO 96 SE (and part of 95 NE) (Hanbury and Stock Green)* 41pp. (Keyworth, Nottingham: Institute of Geological Sciences.)

SO 96 NE	Stoke Prior	R J O Hamblin, 1983. 9pp.
SO 97 NE	Rubery	R J O Hamblin, 1984. 36pp.
SO 97 SE	Bromsgrove	R J O Hamblin, 1983. 24pp.
SP 06 NW	Redditch West	R A Old and A Horton, 1983. 9pp.
SP 06 NE	Redditch East	B C Worssam, 1983. 11pp.
SP 06 SW, SE and parts of 05 NW and NE		
	Astwood Bank and Studley	A A Jackson, 1982. 15pp.
SP 07 NW	Longbridge	R J O Hamblin, 1984. 29pp.
SP 07 NE	Hollywood	R A Old, 1983. 16pp.
SP 07 SW	Alvechurch	R J O Hamblin, 1983. 20pp.
SP 07 SE	Inkford	K Ambrose, 1983. 20pp.
SP 16 and parts of 15 N		
	Henley-in-Arden	P J Strange and K Ambrose, 1982. 25pp.
SP 17 NW	Shirley	R A Old, 1982. 8pp.
SP 17 NE	Solihull and Knowle	R A Old, 1982. 12pp.
SP 17 SW	Tanworth-in-Arden	K Ambrose, 1982. 19pp.
SP 17 SE	Lapworth	K Ambrose, 1982. 12pp.
SP 26 and parts of 25 N		
	Warwick and Hatton	K Ambrose and P J Strange, 1982. 53pp.
SP 27 NW	Berkswell and Balsall Common	R A Old, 1988. 22pp.
SP 27 SW	Wroxall	R A Old, 1988, 19pp.

In addition there are four open file reports dealing with the sand and gravel resources of the district:

Ambrose, K. 1982. *Quaternary deposits of sheets SP 16 and SP 26 W (Henley-in-Arden and Shrewley).* (Keyworth, Nottingham: Institute of Geological Sciences.)

— 1983. *Sheets SP 06 and SO 96 (eastern part) (Redditch and Feckenham). Quaternary deposits with special emphasis on potential resources of sand and gravel. Geological report for DoE: Land Use Planning.* (Keyworth, Nottingham: Institute of Geological Sciences.)

Old, R A. 1982. *Quaternary deposits of sheets SP 17 and SP 27W (Solihull and Balsall Common).* (Keyworth, Nottingham: Institute of Geological Sciences.)

— 1983. *Sheets SO 97 E and SP 07 (Bromsgrove and Alvechurch). Geology with special emphasis on potential resources of sand and gravel.* Geological reports for DoE : Land Use Planning. (Keyworth, Nottingham: Institute of Geological Sciences.)

APPENDIX 3

Borehole catalogue

The most important boreholes in the district are listed below under the appropriate National Grid 1:10 000 quarter sheets. The borehole numbers are those of the BGS 6-inch records system. Boreholes marked * are held 'Commercial in Confidence'. Depths are in metres.

SO 96 NE

1   Sugarbrook No. 1 [9603 6815] Mercia Mudstone 18.7, Bromsgrove Sandstone 287.6. Wills, 1976.

30 Sugarbrook No.3 [9621 6818] Mercia Mudstone 22.9, Bromsgrove Sandstone 398.4

31 Sugarbrook No.4 [9611 6809] Mercia Mudstone 25.0 Bromsgrove Sandstone 375.2

32 Garrington Ltd. (Deritend Stamping Co. Ltd) [9643 6902] Bromsgrove Sandstone 99.7. Wills, 1976.

SO 97 SE

173 Washingstocks No.1 [9598 7323] Bromsgrove Sandstone 93.6. Richardson, 1930; Wills, 1976.

174 Washingstocks No.2 [9582 7318] Bromsgrove Sandstone 121.8, Wildmoor Sandstone 192.3. Richardson, 1930.

175 Washingstocks No.3 [9598 7322] Bromsgrove Sandstone 56.1, Wildmoor Sandstone 164.9, Kidderminster Formation 191.1. Wills, 1976.

176 Burcot No.1 [9847 7162]. Wildmoor Sandstone 61.0, Kidderminster Formation 193.8. Richardson, 1930; Wills, 1976.

177 Burcot No.2 [9848 7165] Wildmoor Sandstone c.61.0, Kidderminster Formation 122.5. Richardson, 1930.

178 Burcot No.3 [9848 7173] Wildmoor Sandstone 66.9, Kidderminster Formation 189.9, Clent Breccia 243.8. Richardson, 1930.

180 Barnsley Hall Hospital [9600 7258]. Bromsgrove Sandstone 20.1, Wildmoor Sandstone 100.0, Kidderminster Formation 106.7. Richardson, 1930.

SP 07 NW

3   Longbridge Pumping station [0072 7755] Kidderminster Formation 90.4, Halesowen Formation 108.1, Wenlock Shale 151.3. Richardson, 1930.

SP 07 SW

55 Brockhill No.1 [0010 7016] Wildmoor Sandstone 42.7, Kidderminster Formation 186.5.

56 Brockhill No.2 [0018 7014]. Wildmoor Sandstone 38.1, Kidderminster Formation 186.5. Wills, 1976.

SP 06 NW

100 Webheath No.1 [0098 6693] Mercia Mudstone c.27, Bromsgrove Sandstone 212.1, Wildmoor Sandstone 346.0, Kidderminster Formation 396.2.

101 Webheath No.3 [0118 6690] Bromsgrove Sandstone 212.1, Wildmoor Sandstone 346.0, Kidderminster Formation 396.2.

SP 06 NE

1   Putcheons Farm (IGS) [0694 6577] Mercia Mudstone 152.4, Bromsgrove Sandstone 165.8. Poole, 1969.

SP 16 NE

66,67   BGS Industrial Minerals Assessment Unit. Cannell and Crofts, 1984.

SP 16 SW

15 Spernall (BGS) [1096 6231] Mercia Mudstone, including Arden Sandstone 50.7.

17 BGS Industrial Minerals Assessment Unit. Cannell and Crofts, 1984.

SP 16 SE

4   Claverdon [1985 6477]. Drift 4.0, Mercia Mudstone 236.5, Bromsgrove Sandstone 250.8. Richardson, 1928.

6-8   BGS Industrial Minerals Assessment Unit. Cannell and Crofts, 1984.

SP 17 NW

176-179   BGS Industrial Minerals Assessment Unit. Cannell and Crofts, 1984.

SP 17 NE

193-199   BGS Industrial Minerals Assessment Unit. Cannell, 1982.

SP 17 SW

186-196   BGS Industrial Minerals Assessment Unit. Cannell and Crofts, 1984

SP 17 SE

64 BGS Industrial Minerals Assessment Unit. Cannell, 1982.

73-79   BGS Industrial Minerals Assessment Unit. Cannell and Crofts, 1984.

SP 25 NW

64 BGS Industrial Minerals Assessment Unit. Cannell and Crofts, 1984.

SP 26 NW

71 Shrewley [2222 6755]. Mercia Mudstone 196.1, Bromsgrove Sandstone 244. Wills, 1976.

74 Rowington [2090 6882] Mercia Mudstone 235.3, Bromsgrove Sandstone 304.8.

82-90   BGS Industrial Minerals Assessment Unit. Cannell and Crofts, 1984.

SP 26 SW

33 Heath End [2321 6085] Mercia Mudstone 136, Bromsgrove Sandstone 182, Enville Group 184.8. Wills, 1976.

49-72   BGS Industrial Minerals Assessment Unit. Cannell and Crofts, 1984.

SP 27 NW

27, 28
30-35   BGS Industrial Minerals Assessment Unit. Cannell, 1982.

SP 27 SW

10*   Meer End (NCB) [2406 7447] Drift and Mercia Mudstone c.147.0, Enville Group c.606, Keele Formation c.851, Halesowen Formation 96.5, Etruria Marl 990.2, Productive Coal Measures 1026.4, Merevale shales 1040.0.

11,12,15   BGS Industrial Minerals Assessment Unit. Cannell, 1982.

14,19,23   BGS Industrial Minerals Assessment Unit. Cannell and Crofts, 1984.

APPENDIX 4

Geological Survey photographs

Copies of these photographs are available for reference in the Library of the British Geological Survey, Keyworth, Nottingham. Colour or black and white prints and 35 mm slides may be supplied at a fixed tariff.

The National Grid References are those of the viewpoints, where known.

CAMBRIAN AND ORDOVICIAN

A 13531–3 Barnt Green Volcanic Formation, Kendal End Farm [0036 7473; 0033 7470; 0033 7470].

A 2006 Lickey Quartzite, showing dip and jointing. Quarry, south end of Cock Green Lane, Rubery.

A 2007 Lickey Quartzite, showing dip and jointing. Quarry at corner of Bristol Road and Leach Green Lane, Rubery.

A 2008 Unconformable junction of Lickey Quartzite and Rubery Sandstone. Quarry at corner of Bristol Road and Leach Green Lane, Rubery.

A 2009 Low anticline in Lickey Quartzite. Leach Heath Quarry, Rubery.

A 2010 Rounded ridge formed of Lickey Quartzite. Rubery Hill, Rubery.

A 2017 Rednal Gorge, a V-shaped valley cut through Lickey Quartzite. Lickey Hill, Rednal.

A 2018 View of North Lickey Hill, Rednal.

A 2019 Gorge cut through Lickey Quartzite, Rednal.

A 2020 Bilberry Hill, a ridge of Lickey Quartzite, Rednal.

A 2021 North Lickey Hill showing wood-covered slope facing south and bare slope facing north.

A 2022–3 Fold in Lickey Quartzite. Quarry, east of Cofton Wood, Lickey Hills, Rednal.

A 13499 Lickey Quartzite at Kendal End, Quarry, view SW [0011 7469].

A 13534 Lickey Quartzite. Old quarry in Barnt Green Road, Rednal [001 751].

PERMIAN

A 13535 Clent Breccia, Beacon Hill, The Lickey [9890 7566].

TRIASSIC

(i) **Kidderminster Formation**

A 13491 Kidderminster Formation at Upper Madeley Farm, view NE [9610 7689].

A 13502 Kidderminster Formation at Marlbrook Gravel Pit, view NNW [982 747].

A 13503 Kidderminster Formation at Marlbrook Gravel Pit, view NE [982 747].

A 13504 Kidderminster Formation at Marlbrook Gravel Pit, view NNE [982 747].

A 13505 Kidderminster Formation sandstone at Shepley Quarry, view NW [9827 7317].

A 13506 Kidderminster Formation sandstone, Shepley Quarry [9827 7317].

A 13507 Kidderminster Formation sandstone, Shepley Quarry, view NE [9843 7310].

(ii) **Wildmoor Sandstone**

A 13757 Wildmoor Sandstone, Vigo Bridge [9860 7133].

(iii) **Bromsgrove Sandstone**

A 2026 Bromsgrove Sandstone in railway-cutting at Bridge No.19 just south of Cofton Tunnel.

A 13500 Bromsgrove Sandstone, Bromsgrove Eastern Bypass, view W [966 708].

A 13501 Bromsgrove Sandstone, Bromsgrove Eastern Bypass, near Townsend Farm [967 719].

A 13508 Bromsgrove Sandstone scarp at Ashborough, Lower Shepley Lane, view SW [9788 7226].

A 13510 Rhythmic sedimentation in high Bromsgrove Sandstone, Aston Fields. [9613 6906].

A 13537 Bromsgrove Sandstone, Finstall Quarry, by Lickey Incline [9763 7025].

A 13758–59 Bromsgrove Sandstone, Burcot Member, Pikes Pool Lane [9836 7138].

A 13773 Bromsgrove Sandstone, Sugarbrook Member, Sugarbrook Road, Aston Fields [9613 6906].

(iv) **Mercia Mudstone**

A 13488 Beaudesert Church (looking NE), Henley-in-Arden [1530 6604].

A 13489 Ridge capped by Arden Sandstone, Henley-in-Arden [157 662].

A 13490 Arden Sandstone showing cross-bedding, Hill Farm, Great Alne [1205 6032].

A 13525 Tanworth-in-Arden Church constructed using Arden Sandstone [114 705].

A 13527 Round Hill and Arden Sandstone in foreground, Great Alne [1089 6223].

A 13530 Basal Beds of Arden Sandstone at type locality, Shrewley Canal cutting [2118 6744].

A 13748 Alvechurch Brick Pit, showing weathered Mercia Mudstone [019 723].

A 13750 Wootton Wawen Church, constructed from Arden Sandstone [154 633].

A 13752 Arden Sandstone, Rowington Canal cutting [202 691].

A 13753 Arden Sandstone, Henley-in-Arden [156 656].

JURASSIC

A 13751 Mary Arden's House, Wilmcote, near Stratford-upon-Avon, constructed from Rugby Limestone [165 583].

QUATERNARY

A 2027–8 Section in sand, clay and gravel drift. Austin Aerodrome, Longbridge.

A 13492   Glacial gravels at Upper Madeley Farm, view NNW.

A 13528–29   Glacial gravels, Pinley Gravel Pit (SE face) [209 662].

VIEWS

A 13498   Newbourne Hill, Alvechurch, view SE [0300 7219].

A 13511   Frankley Beeches from Windmill Hill, Rubery [9736 7785].

A 13512   Rubery from Windmill Hill, near Rubery [9736 7785].

A 13513   The Lickey Hills from Windmill Hill, Rubery [9736 7785].

A 13514   Waseley Hill from Wildmoor, near Money Lane Farm [9663 7661].

A 13515   Waseley Hill and Stock Hill from Wildmoor, near Money Lane Farm [9663 7661].

A 13516   Beacon Hill from Wildmoor, near Money Lane Farm [9663 7661].

A 13526   Morton Bagot Church, constructed from Rugby Limestone and Arden Sandstone [1127 6473].

A 13549   'Rhaetic Escarpment', White House Farm near Aston Cantlow [148 605].

A 13754   Newbourne Hill, Alvechurch [0304 7272].

A 13755   Alvechurch Lodge Farm, Alvechurch [0306 7272].

A 13756   Valley of the River Arrow, Lye Meadows, Alvechurch [0306 7272].

APPENDIX 5

Taxa recorded from the Lias Group

The following list includes specimens collected during the recent survey and by the late Professor T N George, and also notes a number of specimens from the Warwick Museum. The latter were re-examined during the course of the survey, and grateful thanks are extended to the Warwickshire Museum for the loan of the specimens; specimens in parenthesis come from the literature, and have not been re-examined. The specimens were examined by the following:

Plants   : Dr C R Hill (British Museum Natural History)
Fish   : Dr C Patterson and Dr P L Forey (British Museum Natural History)
Crustacea   : Mr S F Morris (British Museum Natural History)
Insects   : Dr P E S Whalley and Dr E A Jarzembowski (British Museum Natural History and Booth Museum, Brighton)
Ammonites   : Prof. D T Donovan (University College, London)
Microfossils   : Dr G Warrington (British Geological Survey)
Other fossils   : Dr H C Ivimey-Cook (British Geological Survey)

PLANTS

Specimens in the Warwickshire Museum

Cycadite pinna	Brown's Wood
Equisetum muensteri	Wilmcote
Cuticles of *Hirmerella airelensis*	Wilmcote
possible leaf of Naiadita	Copt Heath, Knowle
Otozamites bechei	Wilmcote
Pachypteris lanceolata	Wilmcote
?Pagiophyllum peregrinum	Wilmcote

ANTHOZOA

'*Isastraea Tomesii*'	Wilmcote (Phillips, 1871)
'*Septastraea Haimei*'	Wilmcote (Phillips, 1871)
'*Thecosmilia Terquemi*'	Wilmcote (Tomes 1878)

BIVALVIA

Astarte gueuxii
Astarte sp.
Camptonectes subulatus
Cardinia hybrida
C. ovalis
Chlamys cf. textoria
Entolium?
'*Gervillia*' sp.
Gryphaea?

Liostrea hisingeri
L. irregularis
Lucina sp.
Mactromya arenacea
Meleagrinella decussata
Modiolus hillanoides
M. laevis
M. hillanus
M. cf. *liasinus*
M. minimus
'*Mytilus*' *subtilis*
Oxytoma sp.
O. (Palmoxytoma) longicosta
Parallelodon?
Pinna sp.
Placunopsis sp.
Plagiostoma giganteum
Pleuromya?
Plicatula sp.
Protocardia philippiana
Pseudolimea hettangiensis
Pseudopecten?
Pteromya crowcombeia
P. tatei
P. wilkesleyensis

AMMONOIDEA

Alsatites sp.
Caloceras intermedium
C. johnstoni
Psiloceras planorbis
P. plicatulum
Psilophyllites sp.
Schlotheimia sp
Waehneroceras megastoma (Gümbel)

COLEOIDEA

teuthid hooks (Warwickshire Museum)

INSECTA

All specimens are in Warwickshire Museum
Coleoptera elytra (Brown's Wood, Copt Heath and Wilmcote)

Archelcana sp.
Holcoelytum sp.
Orthoptera
Elcanidae
Elcaniid wing

CRUSTACEA

Specimens from the Wilmcote area in the Warwickshire Museum include:

Coleia barrovensis Abundant in certain limestones at several localities and at more than one level, which may include Langport Member beds.

Aeger brodiei

CRINOIDEA

Pentacrinites sp.

ECHINOIDEA

Eodiadema minutum
diadematoid fragments

PISCES

Specimens from Wilmcote in the Warwickshire Museum.

Dapedium cf. *angulifer*
D. cf. *dorsalis*
Furo hastingsiae
Heterolepidotus?
Pholidophorus stricklandi
Ptycholepis minor
Undina barroviensis

REPTILIA

The limestone quarries at Wilmcote have yielded substantial numbers of both ichthyosaur and plesiosaur skeletons, and dispersed bones and teeth. There are collections in the Warwickshire Museum and the British Museum (Natural History), London.

MICROFOSSILS

Bentley Common No.1 Trial pit SO [9725 6570]
MPA 21592: 0.4 m below the Stock Green Limestone

A very sparse organic residue comprising yellow-brown, mostly elongate, fragments of plant tissues and a single specimen of the miospore *Quadraeculina anellaeformis* was recovered from this preparation.

MPA 21591: 0.53 m below Stock Green Limestone

A very sparse organic residue comprising yellow-brown fragments of plant tissues and single specimens of the miospores *Podocarpidites* sp. and *Classopollis torosus* was recovered from this preparation.

MPA 21590: 0.9 m below the Stock Green Limestone
Miospores
Calamospora mesozoica
Deltoidospora sp.
Concavisporites sp.
Acanthotriletes ovalis
Perinopollenites elatoides
indeterminate bisaccate pollen
Classopollis torosus

Organic-walled microplankton
Rhaetogonyaulax rhaetica

FOSSIL INDEX

GENERAL INDEX

BRITISH GEOLOGICAL SURVEY

Keyworth, Nottingham NG12 5GG
(06077) 6111

Murchison House, West Mains Road,
Edinburgh EH9 3LA   031-667 1000

London Information Office, Natural History Museum
Earth Galleries, Exhibition Road, London SW7 2DE
071-589 4090

The full range of Survey publications is available
through the Sales Desks at Keyworth and at Murchison
House, Edinburgh, and in the BGS London Information Office in the Natural History Museum Earth
Galleries. The adjacent bookshop stocks the more
popular books for sale over the counter. Most BGS
books and reports are listed in HMSO's Sectional List
45, and can be bought from HMSO and through
HMSO agents and retailers. Maps are listed in the
BGS Map Catalogue and the Ordnance Survey's Trade
Catalogue, and can be bought from Ordnance Survey
agents as well as from BGS.

*The British Geological Survey carries out the geological survey of
Great Britain and Northern Ireland (the latter as an agency
service for the government of Northern Ireland), and of the
surrounding continental shelf, as well as its basic research
projects. It also undertakes programmes of British technical aid
in geology in developing countries as arranged by the Overseas
Development Administration.*

*The British Geological Survey is a component body of the
Natural Environment Research Council.*

Maps and diagrams in this book use topography based
on Ordnance Survey mapping

HMSO publications are available from:

HMSO Publications Centre
(Mail and telephone orders)
PO Box 276, London SW8 5DT
Telephone orders 071-873 9090
General enquiries 071-873 0011
Queuing system in operation for both numbers

HMSO Bookshops
49 High Holborn, London WC1V 6HB
    071-873 0011 (Counter service only)
258 Broad Street, Birmingham B1 2HE
    021-643 3740
Southey House, 33 Wine Street, Bristol BS1 2BQ
    (0272) 264306
9 Princess Street, Manchester M60 8AS
    061-834 7201
80 Chichester Street, Belfast BT1 4JY
    (0232) 238451
71 Lothian Road, Edinburgh EH3 9AZ
    031-228 4181

HMSO's Accredited Agents
(see Yellow Pages)

And through good booksellers

LANCHESTER LIBRARY